Dieter Enders
Hans-Joachim Gais
Wilhelm Keim
(Editors)

Organic Synthesis via Organometallics (OSM 4)

Dieter Enders
Hans-Joachim Gais
Wilhelm Keim
(Editors)

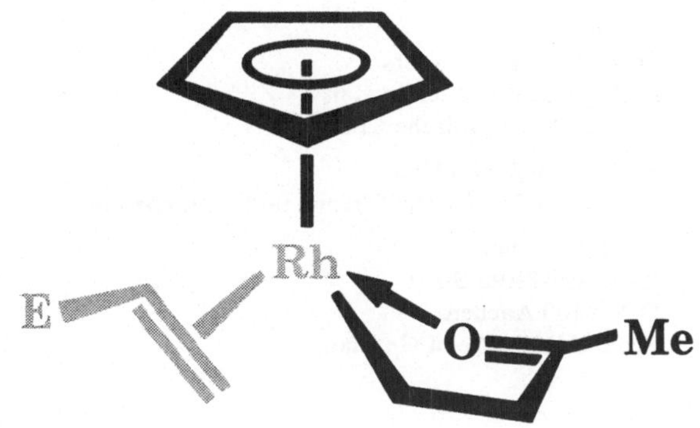

ORGANIC SYNTHESIS VIA ORGANOMETALLICS (OSM 4)

Proceedings of the Fourth Symposium
in Aachen, July 15 to 18, 1992

Editors:
Prof. Dr. Dieter Enders
Prof. Dr. Hans-Joachim Gais
Institut für Organische Chemie

Prof. Dr. Wilhelm Keim
Institut für Technische Chemie und Petrochemie

RWTH Aachen
Professor-Pirlet-Str. 1
D-W 5100 Aachen
Federal Republic of Germany

Vieweg is a subsidiary company of the Bertelsmann Publishing Group International.

ISBN-13: 978-3-528-06481-5 e-ISBN-13: 978-3-322-84062-2
DOI: 10.1007/ 978-3-322-84062-2

Preface

Organometallics play a key role in organic synthesis. Carbon carbon bond formation without main group and transition metal based reagents as well as catalysts is hardly imaginable, and the tremendous success in recent years in the field of stereoselective synthesis of complex biologically active compounds would have been impossible without the advances in organometallic chemistry. From the wealth of carbon carbon bond forming reactions in organotransition metal chemistry many new methods have evolved. This was aided considerably by a deepening of our understanding of the relevant reaction mechanisms. Organometallic chemistry is the bridge par excellence between the traditional fields of inorganic and organic chemistry. It was the intention of the "Volkswagen-Stiftung" to broaden this bridge by starting in 1986 the new interdisciplinary program "Organic Synthesis via Organometallics". From its very beginning this program grew up after 6 years to now more than 60 projects and its encompasses besides its main topic "Organic Synthesis" mechanistic and structural aspects of organometallic chemistry as well. In a series of symposia sponsored by the "Volkswagen-Stiftung", the former of which were held in Hamburg (February 1986), Würzburg (October 1988) and Marburg (July 1990), a forum for intensive discussions and scientific exchange was established. There, scientists participating in the program met with other experts form academia and from industry.

The forth symposium was held in Aachen from July 15 to 18, 1992. Sixteen distinguished and well recognized experts from Germany and abroad had been invited to present recent developments in main group and transition metal mediated asymmetric synthesis, oligomerization, asymmetric catalysis, oxidation and organometallic reaction mechanisms as well as in the synthesis of new organometallics. We feel, that their contributions presented in this volume provide an excellent view on the present state of the art in organic synthesis via organometallics and will stimulate further research in this rapidly advancing field of chemistry. The symposium impressivly demonstrated that organometallic chemistry and especially transition metal catalysis has still much to offer in the future for the achievement of more efficiency in organic synthesis.

Aachen, November 6, 1992

Dieter Enders
Hans-Joachim Gais

Contents

Chiral Aminal Templates.
Diastereo- and enantioselectivity in 1,4 (conjugate) and 1,2 Additions with Organometallic reagents

Alex Alexakis, Richard Sedrani, Nathalie Lensen and Pierre Mangeney

Laboratoire de Chimie des Organoéléments, URA CNRS 473, Université P. et M. Curie,
Tour 44-45 E2, 4 Place Jussieu, 75252 Paris Cedex 05, France

Summary

Chiral diamines, with a *C2* axis of symmetry, form very easily aminals. These aminals act as very efficient chiral controller on both 1,4 (conjugate) and 1,2 additions to prochiral substrates. Steric control or chelation control may account of the high diastereoselectivity observed in these reactions, according to the organometallic reagent and to the solvent.

1. Introduction

During these last years, chiral diamines with a *C2* axis of symmetry emerged as new powerful chiral auxiliaries. They have been used as analytical reagents for the determination of the optical purity of chiral aldehydes [1] and chiral alcohols, thiols and amines [2]. In synthetic organic chemistry, they may be used as catalytic reagents [3] as well as stoichiometric ones [4].

For our part, we have been interested in the formation of aminals, the nitrogen analogues of acetals [5-9]. Thus, not only these diamines may act as protecting group of aldehydes but, also, they act as excellent stereodirecting group for asymmetric synthesis. Compared to acetals [10], the aminal group has the following advantages :

- Its formation is very easy, and, usually, does not need any acid catalyst. Aminals may even be prepared in aqueous media.
- Aminals of ketones are formed in only exceptional cases. Therefore the selectivity for aldehydes is total in compounds having both the ketone and aldehyde functionality.
- Aminals are stable to bases, and their hydrolysis, back to the aldehydes, is done under very mild acidic conditions, without any racemization.

The present lecture points to the comparison of 1,4 addition (conjugate addition) and 1,2 addition of organometallic reagents on substrates bearing a chiral aminal in the proximity of the prochiral center.

2. Conjugate addition

The diastereoselective conjugate addition of organocopper reagent to the modified cinnamate system (**1**) was studied recently by the swedish group [11, 12] :

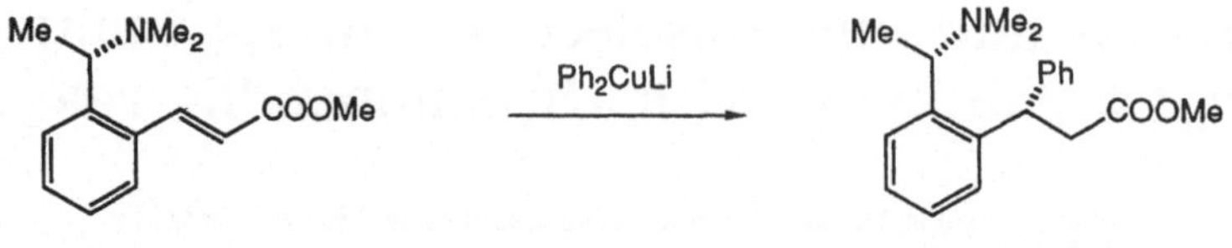

Me NMe$_2$ COOMe — Ph_2CuLi → Me NMe$_2$ Ph COOMe

1 2 d.e. 65%

In collaboration with them, we examined the case of a removable chiral auxiliary placed in the *ortho* position of the aromatic ring. Our first attemps with a variety of acetals (**3a-e**) [13] were completely disapointing :

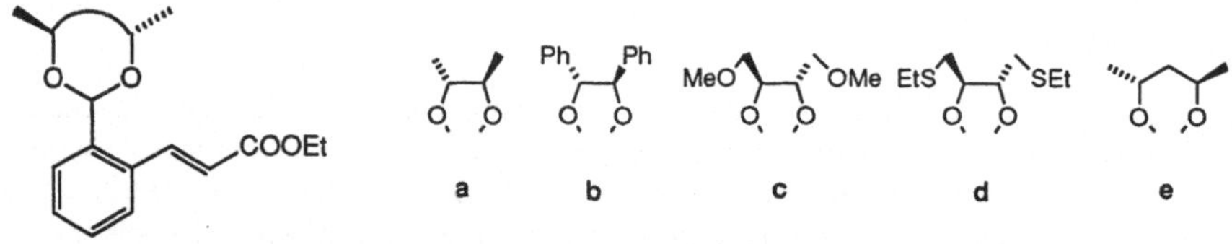

COOEt

3a-e

a b MeO OMe c EtS SEt d e

The diastereomeric excesses varied from 15 to 34%. In fact, as it may be seen in the following Scheme, the stereogenic center of the chiral acetal auxiliary is far away from the prochiral sp2 carbon. A way to bring the stereogenic center closer to the prochiral one, was to shift to oxazolidines, where it is known that the substituent on the nitrogen atom is situated *trans* to the substituents of the adjacent carbons. Thus, it may be considered that the nitrogen becomes the new stereogenic center. However, oxazolidines lack the *C2* symmetry, and during their formation, the carbon 2 becomes a chiral one and its stereochemistry has to be controlled. Fortunately, oxazolidines **6** and **7** form a 93 : 7 mixture, which, by cristallization, afford the pure compound **6** and **7**, respectively from ephedrine and pseudoephedrine.

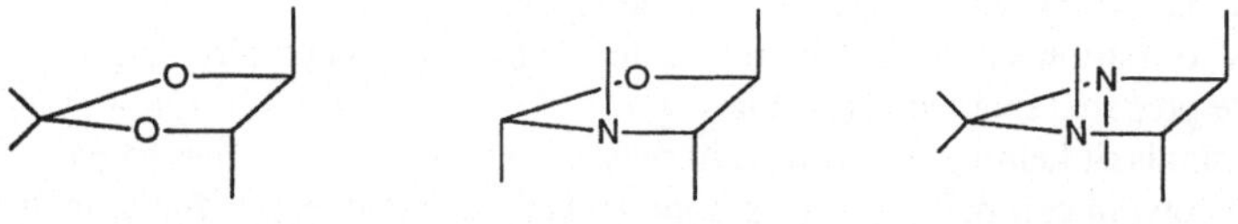

DIOXOLANE OXAZOLIDINE IMIDAZOLIDINE

To come back to the *C2* symmetry both oxygens of the acetal ring may be replaced by nitrogens forming an imidazolidine ring. In such a situation both nitrogens are new stereogenic centers.

Two chiral diamines were mainly used, N,N'-dimethyl cyclohexane-1,2-diamine **4**, easily prepared from commercialy available (R,R)-cyclohexane-1,2-diamine and N,N'-dimethyl-1,2-diphenyl-1,2-ethane diamine **5**, for which we disclosed a new preparation and resolution procedure [5, 6]. The cinnamates **8** and **9** were thus prepared :

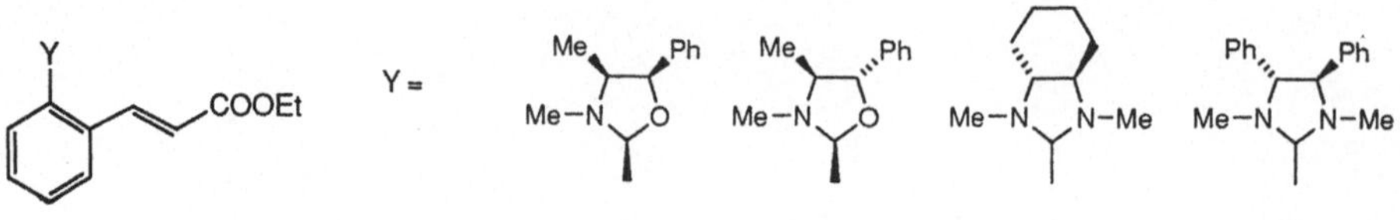

Y COOEt Y =

6 7 8 9

The conjugate addition on these substrates proceeds readily, in Et_2O, with lithium diorganocuprates [14]. The obtained adducts are easily hydrolyzed to the corresponding aldehydes **10a-d** :

Y, COOEt — R_2CuLi / Et_2O → Y, R, *, COOEt — H_3O^+ → CHO, R, *, COOEt **10a-10d**

Table 1. Conjugate addition on cinnamates **6-9**

Cuprate reagent	Product		Oxazolidine **6**	Oxazolidine **7**	Imidazolidine **8**	Imidazolidine **9**
Me_2CuLi	**10a**	yield (%)	51	43	85	57
		e.e. (%)	55 (R)	93 (S)	94 (S)	78 (R)
Bu_2CuLi	**10b**	yield (%)	68	65	90	
		e.e. (%)	62 (R)	70 (S)	95 (S)	
Ph_2CuLi	**10c**	yield (%)	73		84	
		e.e. (%)	98 (S)[a]		96 (R)[a]	
trans Et-CH=CH)$_2$CuLi	**10d**	yield (%)	73		80	
		e.e. (%)	82 (S)[a]		90 (R)[a]	

a This change of sign is only due to the CIP priority rules

As shown in Table 1, the best results are obtained with imidazolidine **8**. However, the most puzzling fact was the inversion of selectivity between aminals **8** and **9**, although they prossess the same absolute configurations. Examination of molecular models seems to indicate that the rigid bicylic structure of aminal **8** favors a conformation where one of the N-Me groups masks the re face of the enoate system. Thus the high diastereoselectivity is probably due to a steric control. In contrast, the imidazolidine ring of **9** is flexible, and it may adopt a different conformation where the complexation of the organometallic reagent by one of the two nitrogens favors the attack on the si face of the enoate system.

In order to confirm these hypotheses it was required to avoid such a complexation and verify that in this case, aminal **9** reverts back to the "normal" stereoselectivity, the same as that of aminal **8**. This task was accomplished by runing the same experiments in the presence of trimethylchlorosilane ((TMSCl), an adjuvent which strongly accelerates the conjugate addition of organocopper reagents [15, 16].

Organocopper reagents form very rapidly π-complexes with enones and enoates. The stereochemistry of this "kinetic" π-complex is primarily dictated by the steric requirements of the substrate. It may equilibrate, then, with a "thermodynamic" π-complex, if another heteroatom is placed in the neighbourhood. By accelerating the conjugate addition with TMSCl the resulting adduct will mainly arise from the "kinetic" π-complex.

This is indeed the case [17]. When the conjugate additions are run in the presence of TMSCl, the stereochemistry of adduct **10** a is not changed with aminal **8**, whereas with aminal **9** the stereochemical course of the reaction is complely inverted. Thus, now, both aminals **8** and **9**, having the same absolute configuration, give the adduct **10** a of the same stereochemistry :

	Me_2CuLi, Et_2O	yield	e.e.
8	without TMSCl	85%	94% (S)
	with TMSCl	62%	78% (S)
9	without TMSCl	57%	78% (R)
	with TMSCl	68%	53% (S)

The aminal auxiliary may also be placed in a closer position to the prochiral center, as in the following fumarate derivatives **11** and **12** [13]. The resulting adducts **13** are hydrolyzed, under mild conditions, to the known optically active aldehydes **14**.

As it may be seen in Table 2, aminal **12** gave only one diastereomer, whereas aminal **11** was not as efficient. In both cases the diastereomeric adducts **13** were stable compounds on which the d.e. could be easily determined by NMR. In contrast to the previous situation, with cinnamates **8** and **9**, here both aminals follow the same stereochemical course. Moreover, the reaction is insensitive to the presence or not of TMSCl, suggesting a steric control of the diastereoselectivity. In the most stable conformation, the attack of the organometallic reagent is believed to occur from the unhindered face, where the N-Me group is pseudo-axial.

Re

MeOOC

Ph

N

N

Ph

Si

R_2CuLi

Table 2. Conjugate additions on fumarates **11** and **12**

Aminal	Cuprate	Yield %	e.e. %	Stereochemistry
11	Me_2CuLi	71	48	R
	Bu_2CuLi	88	69	R
	Bu_2CuLi,TMSCl	70	74	R
12	Me_2CuLi	91	>95	R
	Me_2CuLi,TMSCl	72	>95	R
	Et_2CuLi	87	>95	R
	Bu_2CuLi	85	>95	R
	Ph_2CuLi	82	>95	S

Thus, according to its position and the reaction conditions, the aminal group acts as a steric controler or a coordination one.

3. 1,2 Additions

Conjugate additions of organocopper reagents are considered to occur by a 90° attack on the π system. In contrast, the attack of a nucleophile on a carbonyl group occurs by a 110° angle according to the Bürgi-Dunitz rule. It is clear that a chiral controler would not act in the same manner in these two situations. Accordingly, we prepared analogous substrates where a carbonyl group is the prochiral center instead of the enoate :

Me—N N—Me

15

Ph Ph

Me—N N—Me

16

With such compounds, the choice of the solvent and of the organometallic reagent is much wider. Therefore many of them were tested according to the following Scheme [18] :

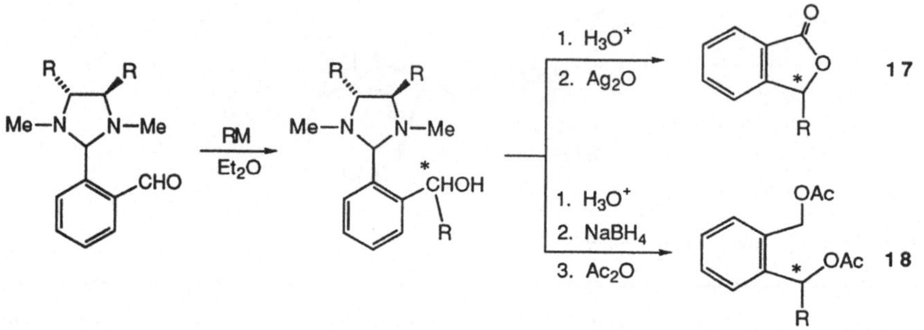

As may be seen in Table 3, the best reagents were organocopper derivatives. This is not surprising, as several reports point to the higher diastereoselectivities obtained with these reagents on reaction with aldehydes [19, 20].

Table 3. Organometallic additions to aldehydes **15** and **16**

Aminal	Organometallic reagent	Overall yield of **18** %	e.e. %	Stereochemistry
15	BuLi	86	0	-
	BuMgBr	72	69	S
	Bu_2CuLi	89	90	S
	Bu_2CuLi,TMSCl	65	95	S
16	BuLi	94	40	R
	BuMgBr	72	68	R
	BuMnBr	69	>99	S
	Bu_2CuLi	74	>99	R
	Bu_2CuLi,TMSCl	52	88	R

Again, both diamines gave contrasting results which may be explained by steric or chelation factors. However, in the case of 1,2 addition, the presence of TMSCl has no marked effects, although some slight differences may be noted.

As compared to the conjugate addition, the diastereoselectivity, here, is completely reversed ! It may be speculated wether cuprate reagents comply to the Bürgi-Dunitz rule on carbonyl addition or if they form again a π-complex by a 90° approach. Whatever the case, it seems that an explanation to this fact may be found on examination of the more stable or more reactive conformation. Cinnamate **8**, for example, cannot adopt an "endo" conformation whereas with aldehyde **15** both conformations "endo" and "exo" are plausible.

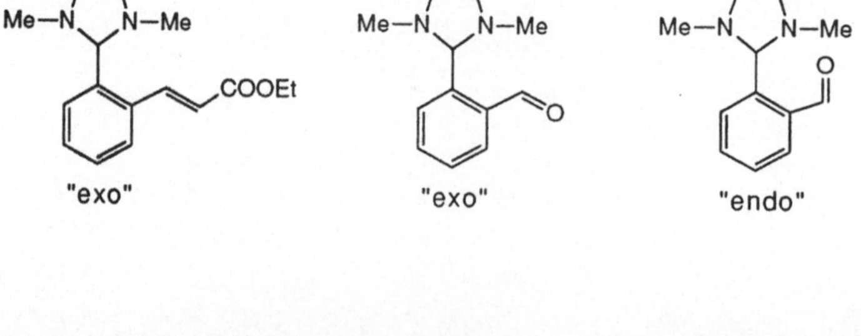

Although slighly less stable, the "endo" conformation seems to be more reactive, because it allows a more favorable approach of the organometallic reagent.

The same comparative study was also made with compounds **19** and **20**, corresponding to cinnamates **11** and **12**. These synthons were very attractive, since they allow a potential entry to the valuable chiral α-hydroxy and α-amino aldehydes :

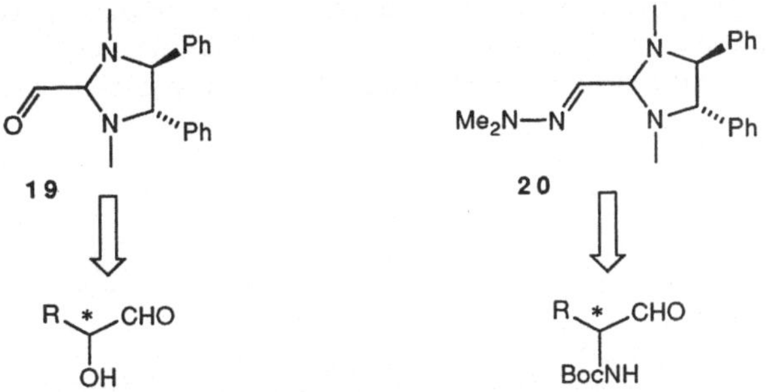

In both cases, the aminal obtained with N,N'-dimethyl-1,2-diamino cyclohexane, could not be prepared efficiently or behaved anomalously. From the variety of reagents added to aldehyde **19**, BuLi in THF gave the highest diastereoselection [21] :

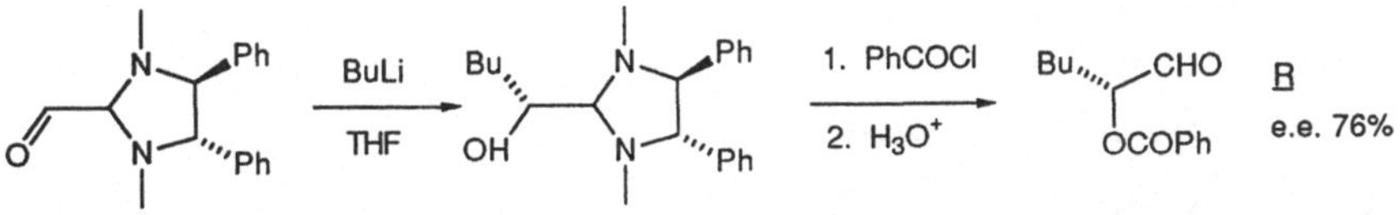

Although the e.e. was not excellent, the diastereomeric aminals could be easily separated by column chromatography.

Our study on hydrazone **20** gave much more interesting results [22]. As seen in Table 4 variation of the solvent and of the organometallic reagent allowed a complete reversal of the diastereoselectivity. This trend is general to various other R group on the organometallic reagent.

Table 4. Reaction of hydrazone 20 with various methyl organometallic reagents.

Organometallic reagent	Solvent	Yield %	d.e. %	Stereochemistry
MeMgBr	THF	0	-	-
"	Et_2O	87	60	R
"	Toluene	85	88	R
BuMgCl	"	83	>99	R
$MeCeCl_2$	THF	67	>99	S
MeLi	Et_2O	72	96	S
"	THF	77	>99	S
BuLi	"	74	>99	S
PhLi	"	80	>99	S

It seems obvious, that in a polar sovent and with a weekly Lewis acidic reagent, such an organolithium reagent, the diastereoselectivity is mainly controlled by steric factors. In the case of the conjugate addition on fumarate **12** we invoked the same arguments. And, indeed, in both cases (1,2 addition on hydrazone **20** and 1,4 addition on fumarate **12**) the stereochemistry of the newly created stereogenic center is the same. On the other hand, a Lewis acidic organometallic reagent, such as a Grignard reagent, in a weekly polar solvent (toluene), favors a chelation process which inverts the diastereoselectivity. It may be easily seen on models, that an "endo" conformation allows such a chelation ; 1°) by the hydrazone nitrogen and 2°) by the imidazolidine ring nitrogen having the pseudo-axial lone pair.

Re Si Me Me N N N N Ph Ph RLi

Si Re Me R X Me N N Mg N N Ph Ph RMgX

Thus, from the same hydrazone **20** it is possible to obtain both enantiomers of the desired chiral α-amino aldehydes. These latter usefull synthons are easily obtained after cleavage of the N-N bond of the hydrazine, followed by protection of the free amine functionality and hydrolysis of the aminal to the aldehyde.

R Me_2N—NH N N Ph Ph H_2 / Raney Ni R NH_2 N N Ph Ph 1. $Boc)_2O$ 2. HCl 2% R * CHO BocNH

Noteworthy, in this process, are the ultrasound assisted cleavage of the N-N bond [23], a notable improvement of a classical method, and the mild, non-racemizing, conditions of hydrolysis of the aminal chiral auxiliary (and protecting group).

4. Conclusion

The above examples are illustrative of the synthetic uses of chiral diamines in asymmetric synthesis. The formation of aminals, as well as their hydrolysis back to the aldehyde, are simple and non-racemizing processes. These aminal chiral controllers are highly efficient and this fact may be ascribed either to non-bonding interactions or to coordination of the organometallic reagent by the lone pair of the nitrogen atom. Both theses situations have been encountered, and by the appropriate choice of the organometallic reagent and of the solvent it is possible to attain very high degree of diastereoselectivity and, hence, enantioselectivity.The comparison of the 1,4 and 1,2 additions shows the same trend of diastereocontrol although the conformational changes, in the case of aldehydes, may reverse the site-selectivity.

References

[1] D. Cuvinot, P. Mangeney, A. Alexakis, J.F. Normant, J. Org. Chem. 54 (1989) 2420

[2] A. Alexakis, S. Mutti, P. Mangeney, J. Org. Chem. 57 (1992) 1224

[3] See for example : E.N. Jacobsen, W. Zhang , A.R. Muci, J.R. Ecker, L. Deng, J. Am. Chem. Soc. 113 (1991) 7063 and references cited therein

[4] E.J. Corey, Pure Appl. Chem. 62 (1990) 1209

[5] P. Mangeney, T. Tejero, A. Alexakis, F. Grojean, J.F. Normant, Synthesis (1988) 255

[6] P. Mangeney, F. Grojean, A. Alexakis, J.F. Normant, Tetrahedron Lett. 29 (1988) 2675 and 2677

[7] M. Commerçon, P. Mangeney, T. Tejero, A. Alexakis, Tetrahedron Asymmetry 1 (1990) 287

[8] R. Gosmini, P. Mangeney, A. Alexakis, M. Commerçon, J.F. Normant, Synlett (1991) 111

[9] P. Mangeney, R. Gosmini, A. Alexakis, Tetrahedron Lett. 32 (1991) 3981

[10] A. Alexakis, P. Mangeney, Tetrahedron Asymmetry 1 (1990) 477

[11] C. Ullenius, B. Christenson, Pure Appl. Chem. 60 (1988) 57

[12] B. Christenson, T. Olsson, C. Ullenius, Tetrahedron 45 (1989) 523

[13] R. Sedrani, Thèse de Doctorat de l'Université P. et M. Curie, Paris, 1990

[14] A. Alexakis, R. Sedrani, P. Mangeney, J.F. Normant, Tetrahedron Lett. 29 (1988) 2951

[15] E.J. Corey, N.W. Boaz, Tetrahedron Lett. 26 (1985) 6015

[16] A. Alexakis, J. Berlan, Y. Besace, Tetrahedron Lett. 27 (1986) 2143

[17] A. Alexakis, R. Sedrani, P. Mangeney, Tetrahedron Lett. 31 (1990) 345

[18] A. Alexakis, R. Sedrani, P. Mangeney, Tetrahedron Asymmetry 1 (1990) 238

[19] W.C. Still, J.A. Schneider, Tetrahedron Lett. 21 (1980) 1035

[20] T. Kunz, H.U. Reissig, Ang. Chem. Int. Ed. Engl. 27 (1988) 268

[21] N. Lensen, Thèse de Doctorat de l'Université P. et M. Curie, Paris, 1992

[22] A. Alexakis, N. Lensen, P. Mangeney, Tetrahedron Lett. 32 (1991) 1171

[23] A. Alexakis, N. Lensen, P. Mangeney, Synlett (1991) 625.

New Catalytic Asymmetric Carbon-Carbon Bond-Forming Reactions by Palladium and by Lanthanum

Masakatsu Shibasaki

Faculty of Pharmaceutical Sciences, University of Tokyo, Hongo, Bunkyo-ku, Tokyo 113, Japan

Summary

In 1989 we succeeded in demonstrating the first example of an asymmetric Heck reaction. Since then, an asymmetric Heck reaction has been successfully applied to catalytic asymmetric syntheses of *cis*-decalin derivatives, *cis*-hydrindan derivatives and polycyclopentanoids with enantiomeric excesses ranging from 80% to 92%. Formation of 16-electron Pd^+ intermediates has been found to be essential to obtain products with high ees. Furthermore, in 1991 we succeeded in demonstrating the first example of a catalytic asymmetric nitroaldol reaction utilizing an optically active lanthanum complex, which gave a nitroaldol in up to 95% ee.

1 Introduction

The synthesis of optically active compounds is an extremely important undertaking and a formidable challenge to the synthetic chemist, because enantiomer recognition plays an important role in biological activity. Of the various ways to induce enantioselectivity in chemical reactions, the most efficient is by means of a chiral catalyst, where a small amount of chiral material can transmit chirality information to a large amount of substrate. Although a lot of successful studies on catalytic asymmetric epoxidation [1], hydrogenation [2], and their synthetic application have been reported, there are only a few excellent catalytic asymmetric C-C bond-forming reactions [3]. Herein, we describe two catalytic asymmetric carbon-carbon bond-forming reactions by making the best use of the characteristics of elements such as palladium and lanthanum.

2 Asymmetric Heck reaction

The Heck reaction has been recognized to be quite useful for catalytic carbon-carbon bond-forming reactions [4]. Until 1989, however, the Heck reaction had never been used to catalyze the formation of an

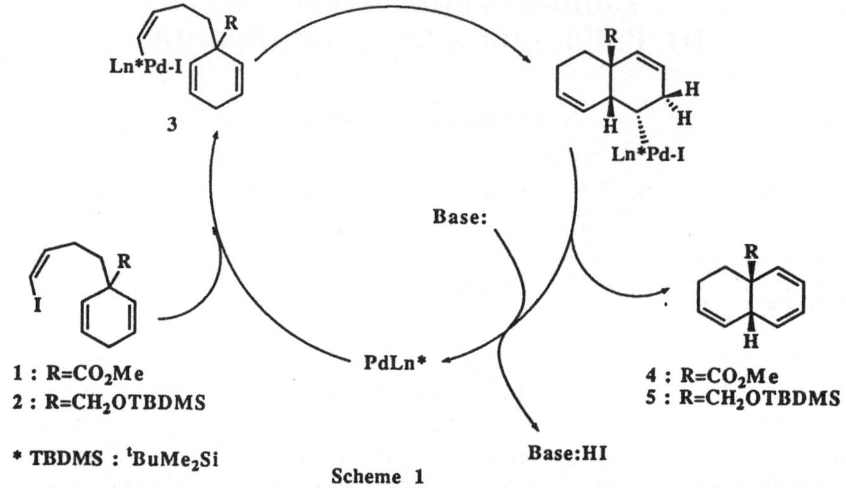

Ln*Pd-I

3

R

H

H

H

Ln*Pd-I

Base:

I

R

1 : R=CO_2Me
2 : R=CH_2OTBDMS

* TBDMS : ^{t}BuMe_2Si

PdLn*

Base:HI

4 : R=CO_2Me
5 : R=CH_2OTBDMS

Scheme 1

HO — H; 1) TsCl, Py 2) NaI, acetone reflux, 55%; I — H; I_2, CuI, ^{n}Bu_4NI, Na_2CO_3, DMF, 60%; I — I

NCO_2K ‖ NCO_2K, AcOH, MeOH, 61%

CO_2H; 1) Na, EtOH, NH_3 2) CH_2N_2, 88%; CO_2Me; LDA, THF; 85%

I; OTBDMS; TBDMSCl, NEt_3, DMAP; 2; OH; $LiBH_4$, Et_2O, 69%; CO_2Me; 1

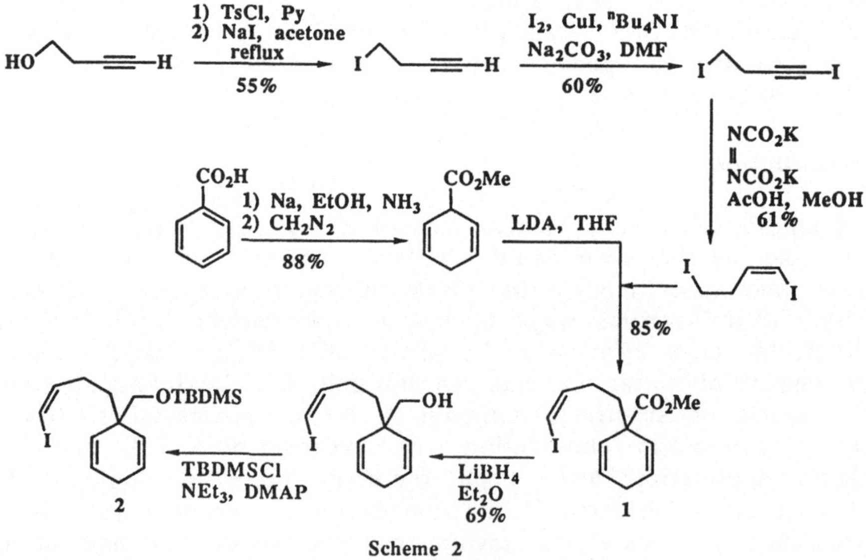

Scheme 2

asymmetric carbon-carbon bond. This reaction is not an obvious choice for asymmetric catalysis, probably due to the fact that in its most general form it does not generate a chiral center and, in general, it does not require a phosphorus ligand. However, including a prochiral center in the substrate provides the opportunity to set two chiral centers in one step as shown in Scheme 1. We therefore selected the prochiral alkenyl iodide **1** as well as the corresponding alkenyl triflate **8** as a starting substrate for an asymmetric Heck reaction, in order to produce the optically active *cis*-decalin derivative **4**. Since no definite mechanistic information on the Heck reaction, concerning whether or not partial dissociation of ligands occurs at the stage of olefin coordination to an alkenylpalladium iodide **3** is available, application of the Heck reaction to a catalytic asymmetric synthesis is a quite challenging research field. The requisite prochiral alkenyl iodides **1** and **2** were effectively prepared as shown in Scheme 2. First of all, we examined the Heck-type reaction of **1** utilizing an achiral phosphorus ligand. With the aim of application to an asymmetric synthesis, the reaction using $Pd(OAc)_2$ as a catalyst and (diphenylphosphino)ethane (DIPHOS) as a ligand was carefully investigated. When *N,N*-diisopropylethylamine or sodium acetate was used as a base, the reactions were very slow in all the solvents examined even at 100 °C, providing the cyclized product as a mixture of some regioisomers in low yields. On the other hand, the use of Ag_2CO_3 according to the procedure developed by Hallberg and by Overman [5] afforded the cyclized product **4** in 68% yield as a single isomer. The result is shown in Table 1.

Table 1. Synthesis of (±)-4 by $Pd(OAc)_2$ and (Diphenylphosphino)ethane (DIPHOS)

1 —— 5 mol % $Pd(OAc)_2$, 5.5 mol % Ph_2P⌒PPh_2 ——→ (±)-4

run	base	solvent	temp,°C	time, h	S.M. recov[d], %	yield, %
1	iPr_2EtN	DMF	70-100	164	56	16[a]
2	iPr_2EtN	THF	reflux	65	66	16[a]
3	iPr_2EtN	CH_3CN	reflux	117	62	11[a]
4	iPr_2EtN	toluene	100	66	45	37[a]
5	AcONa	toluene	100	94	34	12[a,b]
6	Ag_2CO_3	CH_3CN	60	5	3	68[c]

a Obtained as a mixture of some regioisomers (based on ^{1}NMR).

b Methyl benzoate was formed in 45% yield.

c (±)-4 was obtained as a single isomer.

d S.M. = starting material

Having established a Heck-type cyclization of **1**, we then turned our attention to an asymmetric Heck reaction. After many experiments utilizing $Pd(OAc)_2$, various optically active bidentate phosphorus ligands and Ag_2CO_3 in a variety of solvents, it was found that the use of BINAP (1.1 molar equiv per Pd) in 1-methyl-2-pyrrolidinone (NMP) gave the modest enantiomeric excess (33%) in 54% yield [6]. It was further observed that under the conditions described above the reaction proceeded by several catalyst species including the only solvent-palladium complex. This drawback could be overcome by using 3-5 equivalents of BINAP per palladium, giving **4** of 50-60% ee. Finally, the nearly same enantiomeric excess was obtained by the use of $PdCl_2$-BINAP complex in spite of the Pd-BINAP ratio (1:1). Next, in order to clarify the effect of Ag_2CO_3 on ee, the asymmetric Heck reaction utilizing Et_3N as a base was carried out. And it was found that the enantiomeric excess of the product was only 0.4% (low chemical yield), revealing that Ag_2CO_3 played a key role in the enantioselectivity as well as the chemical yield. These results are understandable by assuming that the Heck reaction proceeds via the 16-electron Pd^+ intermediate **6** in the presence of Ag_2CO_3, while the reaction proceeds via the intermediate with a partially dissociated bidentate phosphorus ligand **7** in the absence of Ag_2CO_3 as shown in Scheme 3. Having observed that addition of Ag_2CO_3 is essential to obtain the product of relatively high ee, we then investigated the effects of other various silver salts on the asymmetric induction. The results are summarized in Table 2.

Table 2. Catalytic Asymmetric Cyclization of the Prochiral Iodides **1** and **2** Using $Cl_2Pd(R)$-BINAP in the Presence of Silver Salt

run	substrate	silver salt	method[a]	time, h	SM recov., %	product	yield, %	ee, %
1	**1**	Ag_2CO_3	A	210	-	**4**	65	58
2	**1**	Ag_2O	A	48	11	**4**	54	63
3	**1**	Ag_3PO_4	B	188	12	**4**	48	69
4	**1**	Ag_2SO_4	B	188	85	**4**	11	53
5	**1**	$AgBF_4$	B	252	-	**4**	27	26
6	**1**	$AgNO_3$	B	134	-	**4**	39	27
7	**1**	$AgClO_4$	B	230	-	**4**	33	29
8	**1**	AgOTf	B	208	-	**4**	31	23
9	**1**	AgOAc	B	61	-	**4**	70	6
10	**1**	Ag_2CO_3	C	162	-	**4**	58	70
11	**1**	Ag_2O	C	90	-	**4**	55	67
12	**2**	Ag_2CO_3	A	156	-	**5**	58	68
13	**2**	Ag_2O	A	66	-	**5**	59	77
14	**2**	Ag_3PO_4	B	84	-	**5**	67	80

a mehtod A : The alkenyl iodide was treated with $Cl_2Pd(R)$-BINAP (10 mol %) and the silver salt (2 molar equiv) in NMP at 60 °C ; method B : method A + 2.2 molar equiv of $CaCO_3$; method C : The Pd^0 catalyst generated in situ by reaction of $Cl_2Pd(R)$-BINAP (5 mol %) with cyclohexene (10 mol %) in the presence of (*R*)-BINAP (5 mol %) and silver salt (10 mol %) (NMP solvent, 60 °C, 1h) was added to the alkenyl iodide and Ag_2CO_3 (1.9 molar equiv) in NMP, and the reaction mixture was stirred at 60 °C.

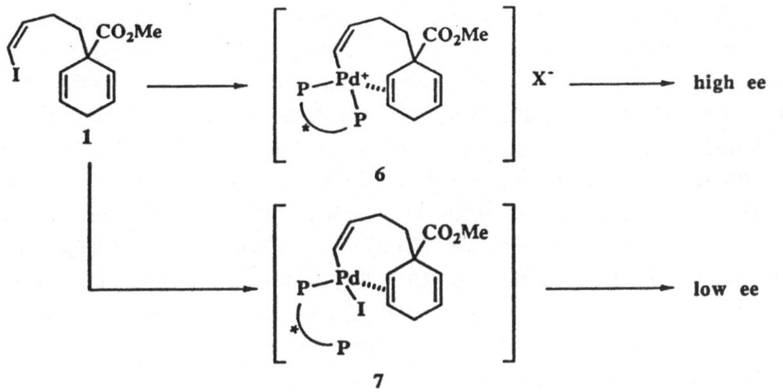

CO_2Me I 1 → [6: P—Pd⁺, P, *, CO_2Me] X^- → high ee

1 → [7: P—Pd, I, P, *, CO_2Me] → low ee

Scheme 3

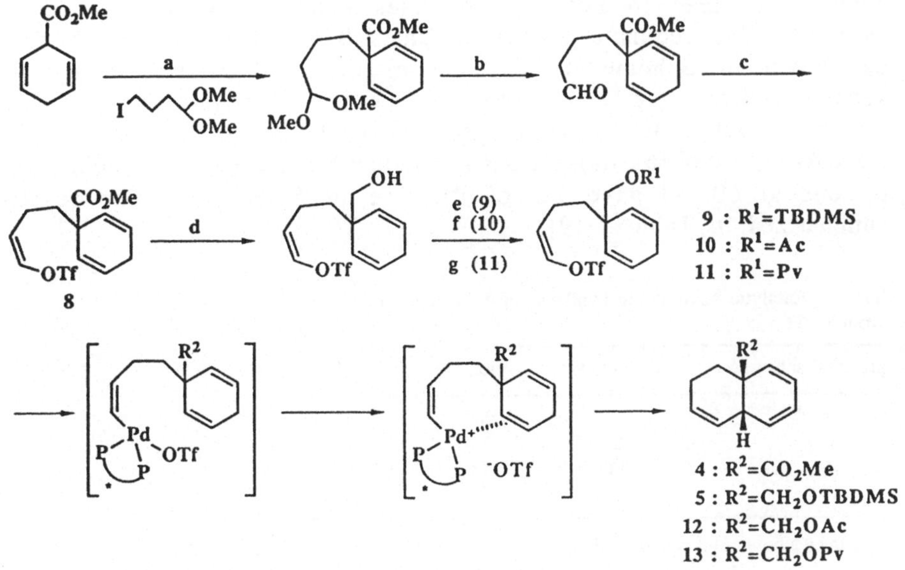

CO_2Me —a ($I\diagup\diagdown\diagup$OMe, OMe)→ CO_2Me, MeO, OMe —b→ CO_2Me, CHO —c→

8 (CO_2Me, OTf) —d→ OH, OTf —e (9), f (10), g (11)→ OR^1, OTf

9 : R^1=TBDMS
10 : R^1=Ac
11 : R^1=Pv

→ [R^2, Pd, P, P, OTf, *] → [R^2, Pd⁺, P, P, $^-$OTf, *] → R^2, H

4 : R^2=CO_2Me
5 : R^2=$CH_2OTBDMS$
12 : R^2=CH_2OAc
13 : R^2=CH_2OPv

a. LDA, THF, 0 °C (76%); b. TsOH, acetone, r.t. (100%); c. Tf_2O, 2,6-di-*tert*-butylpyridine, 1,2-dichloroethane, refulux (63%); d. LiAlH4, Et_2O, -78 °C (85%); e. TBDMSCl, imidazole, DMF (95%); f. Ac_2O, pyridine, DMAP, CH_2Cl_2 (96%); g. PvCl, pyridine, DMAP, CH_2Cl_2 (98%)

Scheme 4

It is noteworthy that the use of Ag_3PO_4 gives the best enantiomeric excess and, in particular, the prochiral substrate **2** is converted to **5** of 80% ee in 67% yield. When the asymmetric Heck reaction was carried out utilizing AgOAc as a base, the reaction was found to proceed rather rapidly, indicating that the reaction moved on via the Pd^+ intermediate. The enantiomeric excess of the product, however, was only 6%. At the moment it is very difficult to understand the effects of silver salts on the asymmetric induction. We are assuming that the identity of the counter anion (X^-) influences on the asymmetric induction. That is, counter anions such as AcO^- appear to make tight ion pairs with Pd^+, thus giving a product with low ee through an intermediate with a partially dissociated bidentate phosphorus ligand. On the other hand, counter anions generated from silver salts such as Ag_3PO_4, Ag_2O and Ag_2CO_3 seem not to make tight ion pairs with Pd^+, thus producing a product with relatively high ee via an intermediate such as **6** [7]. Having recognized the essential role of a Pd^+ intermediate in an asymmetric Heck reaction, we then turned our attention to a catalytic asymmetric synthesis utilizing prochiral alkenyl triflates, which were expected to lead to Pd^+ intermediates even in the absence of silver salts [8]. The requisite alkenyl triflates **8**, **9**, **10** and **11** were prepared as shown in Scheme 4, and the major *cis*-isomers were readily separated from the *trans*-isomers (Z:E=5:1) by silica gel column chromatography. We were pleased to find that treatment of **11** with $Pd(OAc)_2$ (5 mol %), (*R*)-BINAP (10 mol %) and K_2CO_3 (2 molar equiv) in toluene at 60 °C gave **13** of 91% ee in 60% yield. The results are summarized in Table 3 [9].

Table 3. Catalytic Asymmetric Synthesis of **4**, **5**, **12** and **13** from **8** - **11**

prochiral substrate	product	yield, %	ee, %
8	4	54	91[a]
9	5	35	92
10	12	44	89
11	13	60	91

a A trace amount of **8** was recovered.

Next we became interested in a catalytic asymmetric synthesis of more functionalized decalin derivatives, which would find immediate application in the synthesis of bioactive natural products. The prochiral alkenyl iodides **15** and **16**, again readily prepared from methyl dihydrobenzoate, was treated with $PdCl_2$[(*R*)-BINAP] (10 mol %), Ag_3PO_4 (2 molar equiv) and $CaCO_3$ (2 molar equiv) in NMP at 60 °C for 41 hr to give the expected product **17** with 83% ee in 63% yield

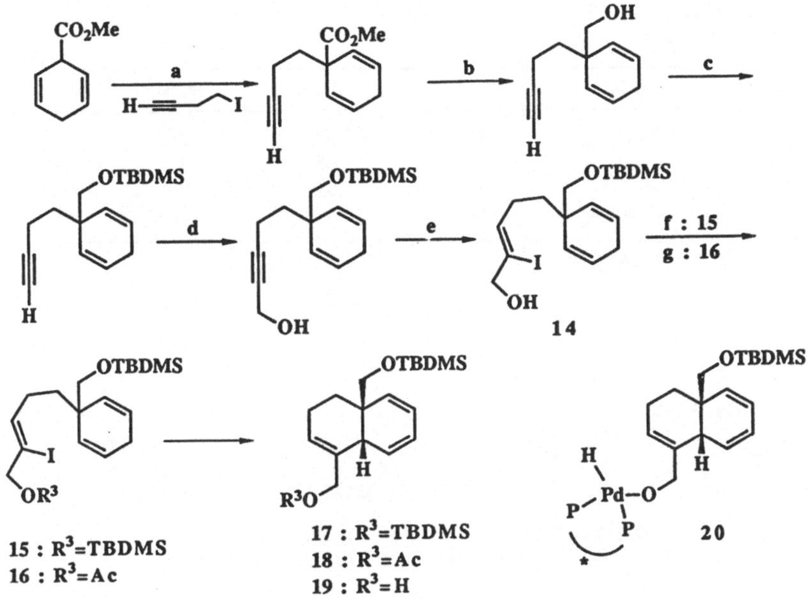

a. LDA, THF-HMPA, -78 °C then 0 °C (74%); b, $LiAlH_4$, Et_2O, -40 °C (100%); c. TBDMSCl, imidazole, DMF (100%); d. i) EtMgBr, THF, then DMF, -30 °C ii) NaBH4, MeOH, 0 °C (83%, 2 steps); e. *n*-BuLi, DIBAH, Et_2O, 35 °C then I_2, -78 °C (61%); f. TBDMSCl, imidazole, DMF (89%); g. Ac_2O, pyridine, DMAP, CH_2Cl_2 (85%).

Scheme 5

accompanied with the allylic alcohol **19** of 92% ee (35% yield). In order to understand the mechanism of the formation of **19** with 92% ee, the following experiments were carried out. First, the prochiral allylic alcohol **14** underwent cyclization under the similar conditions as described above to furnish **19** with lower ee (71% ee), revealing that **19** (92% ee) was not produced from **14**. Second, treatment of **17** with $PdCl_2$[(*R*)-BINAP], Ag_3PO_4, $CaCO_3$ and 1.5 equiv of *n*-Bu_4NOAc in NMP at 60 °C for 44 hr afforded none of **18**, ruling out a possibility of the formation of the π-allylpalladium complex from **17**. These results appear to suggest that transmetalation between silicon and $(BINAP)Pd^{+}H$ [**17**→**20**] plays a key role in the above mentioned kinetic resolution. Next, a catalytic asymmetric cyclization of the prochiral alkenyl iodide **16** was investigated, giving **18** with 87% ee in 67% yield. It is interesting to note that there is a similar tendency for silver salts to effect on enantiomeric excess as observed previously. The results are summarized in Table 4[9].

Furthermore, an asymmetric Heck reaction was found to be useful for the catalytic asymmetric synthesis of *cis*-hydrindan derivatives. At the outset, we carried out an examination of a catalytic asymmetric

Table 4. Catalytic Asymmetric Synthesis of 18 from 16 under Various Conditions[a]

silver salt	$CaCO_3$	time, hr	yield, %	ee, %	SM recov., %
Ag_3PO_4	2.2 mol eq	100	67	87	—
Ag_2O	—	90	68	70	—
Ag_2CO_3	—	140	33	65	—
AgOAc	2.2 mol eq	191	47	23	48

a 10 mol % $PdCl_2$[(*R*)-BINAP], NMP

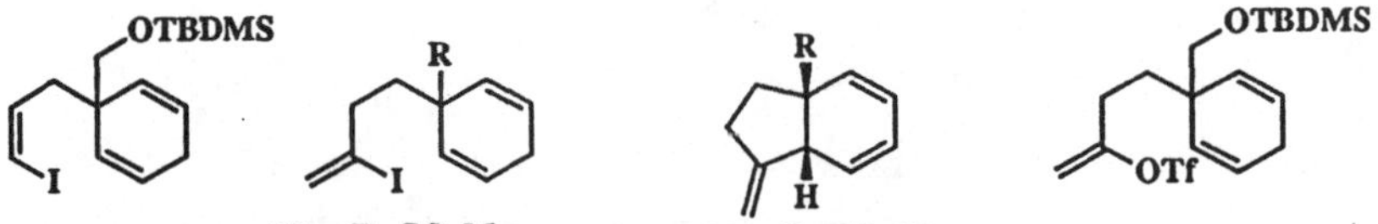

21

22 : R=CO_2Me
23 : R=$CH_2OTBDMS$

24 : R=CO_2Me
25 : R=$CH_2OTBDMS$

26

Scheme 6

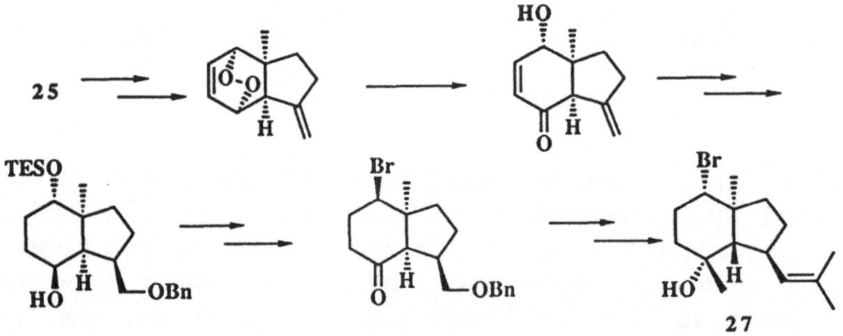

27

Scheme 7

synthesis utilizing the prochiral alkenyl iodide **21**. However, it became clear that **21** did not produce the cyclized product with high ee under a variety of reaction conditions. Thus, we next investigated a catalytic asymmetric synthesis using the prochiral alkenyl iodide **22**. The best result was obtained as described in the following. $PdCl_2$[(*R*)-BINAP] (10 mol %) was reduced by treatment with cyclohexene (20 mol %), Ag_3PO_4 (20 mol %) and $CaCO_3$ (22 mol %) in NMP at 60 °C for 1 hr, and the mixture was added to a suspention of **22**, Ag_3PO_4 (1.8 molar equiv) and $CaCO_3$ (2.0 molar equiv) in NMP. The resulting suspension was stirred at 40 °C for 138 hr, furnishing the *cis*-hydrindan **24** with 86% ee in 72% yield. Likewise, the corresponding prochiral alkenyl triflate **26** was transformed into **25** of 73% ee in 63% yield [10]. The *cis*-hydrindan derivative **25** was successfully converted to oppositpl **27**, a marine natural product, as shown in Scheme 7.

Finally, a catalytic asymmetric synthesis of capnellenols is described. Capnellenols are sesquiterpene alcohols **28-33**, isolated from sun-dried colonies of the soft coral *Capnella imbricata.* These substances appear to have protective roles against fish predation and invasion by microorganisms, larvae, and algae. In 1986, we compleated the first total synthesis of (±)-**28**, (±)-**30** and (±)-**33** using **44** and **45** as key intermediates [11]. We reasoned that treatment of **34** with a palladium catalyst bearing a chiral ligand in the presence of a silver salt and some oxygen nucleophile (ROH) would regioselectively afford the optically active bicyclic compound **36**, a potential intermediate for **30** and **33**, via the π-allyl intermediate **35**. The regiochemistry of **36** was expected to be controlled by steric factors. First, with the aim of application to an asymmetric synthesis, reactions utilizing DIPHOS as a ligand were investigated, and it was found that exposure of **34** to $Pd(OAc)_2$ (5.8 mol %), DIPHOS (5.7 mol %) and tetrabutylammonium acetate (1.72 equiv) in CH_3CN (60 °C, 112 hr) gave **36** (R=Ac, 52%) in a highly stereo- and regiocontrolled manner. The reaction did not proceed in the absence of tetrabutylammonium acetate. This reaction was next applied to a catalytic asymmetric synthesis. However, addition of a silver salt to the reaction medium was found to cause the decomposition of **34**, probably owing to the presence of the cyclopentadiene moiety. For this reaction, **36** (R=Ac) was obtained with only a low ee. That is, treatment of **34** with $[Pd(allyl)Cl]_2$ (10 mol %), (*R*,*R*)-CHIRAPHOS (10 mol %), and tetrabutylammonium acetate (2.9

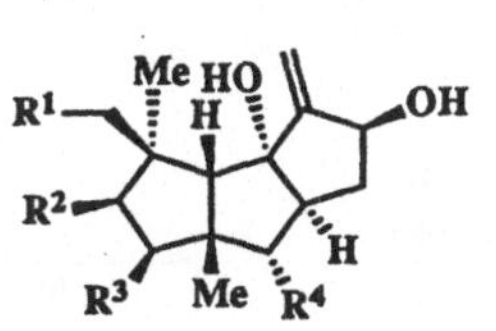

Me HO H OH R^1 R^2 R^3 Me R^4 H

28 : $R^1=R^2=R^3=R^4=H$
29 : $R^1=R^3=R^4=H$, $R^2=OH$
30 : $R^1=R^2=R^4=H$, $R^3=OH$
31 : $R^1=R^2=R^3=H$, $R^4=OH$
32 : $R^1=R^3=H$, $R^2=R^4=OH$
33 : $R^2=R^4=H$, $R^1=R^3=OH$

Scheme 8

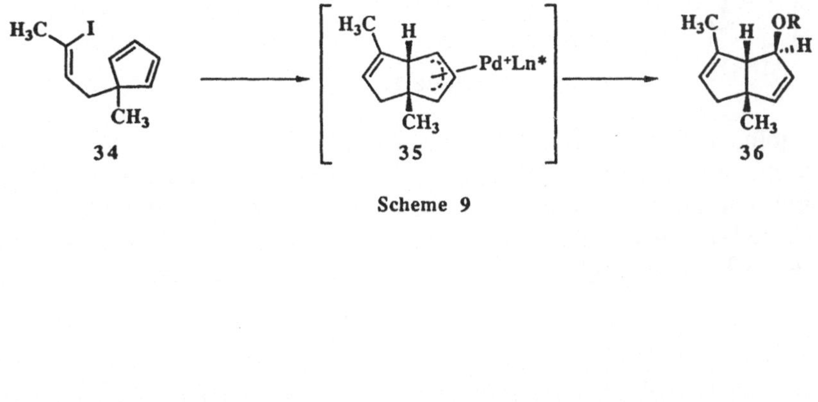

H_3C I CH_3 **34** → [H_3C H Pd^+Ln^* CH_3 **35**] → H_3C H OR H CH_3 **36**

Scheme 9

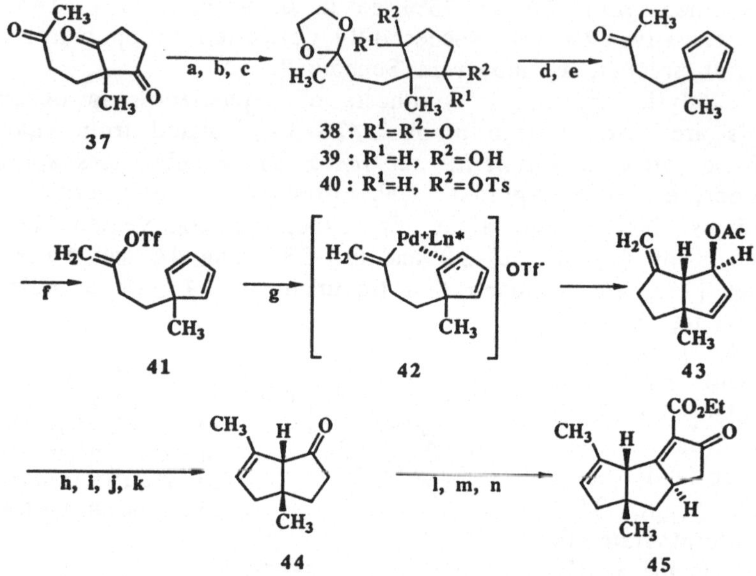

a : $(TMSOCH_2)_2$, TMSOTf (77%); b : $NaBH_4$; c : TsCl, DMAP, pyridine; d : DBU (88%); e : TsOH, acetone (88%); f : LDA, Tf_2NPh (71%); g : $Pd(OAc)_2$, (*S*)-BINAP, Bu_4NOAc, DMSO(89%); h : NaOMe (85%); i : PDC,MS3A (95%); j : CuBr, Red-Al, *s*-BuOH, THF (83%); k : DBU (90%); l : LDA, $ICH_2C(OMe){=}CHCO_2Et$ (85%); m : 30% $HClO_4$, Et_2O (90%); n : NaOEt, EtOH (85%).

Scheme 10

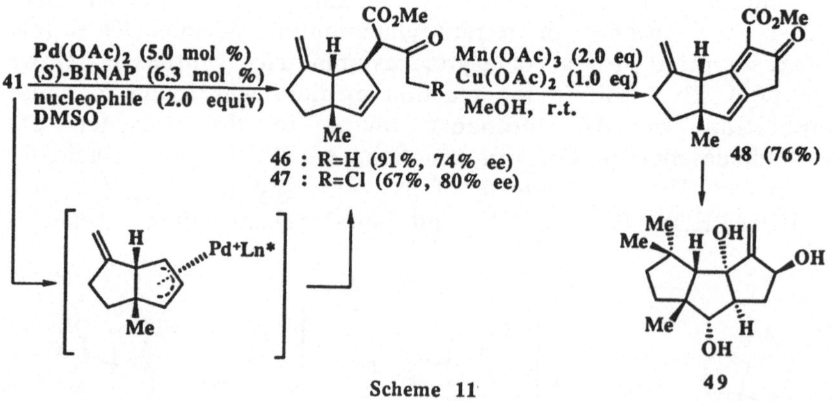

Scheme 11

equiv) in toluene (60 °C, 144 hr) provided **36** (R=Ac) with 20% ee (61% yield). Next, we undertook a catalytic asymmetric cyclization utilizing the alkenyl triflate **41**, which was expected to produce the 16-electron Pd^+ intermediate **42** efficiently even in the absence of a silver salt, leading to **43** with high ee. The prochiral alkenyl triflate **41** was readily prepared in 42% overall yield starting with **37** as shown in Scheme 10. We were pleased to find that treatment of **41** with $Pd(OAc)_2$ (1.7 mol %), (*S*)-BINAP (2.1 mol %), and tetrabutylammonium acetate (1.7 equiv) in DMSO at 20 °C for 2.5 hr produced **43** in 80% ee (89% yield). The cyclized product **43** was then converted to **44** : $[\alpha]^{20}_D$ +532 ° (*c* 0.85, $CHCl_3$) (80% ee), the key intermediate for **30** and **33**, in a four-step process (60% overall yield). Furthermore, **44** was transformed into the ABC ring system **45** : $[\alpha]^{21}_D$ +530 ° (*c* 1.00, $CHCl_3$) (80% ee) in a three-step process (65% overall yield) [12]. It was also found that treatment of **41** with $Pd(OAc)_2$ (5.0 mol %), (*S*)-BINAP (6.3 mol %) and a C-centered nucleophile in DMSO gave versatile bicyclic products such as **46** and **47** in one-pot as shown in Scheme 11. The compound **46** was already converted to **48**, which appears to be a potential intermediate for capnellene **49**. As we have described herein, an asymmetric Heck reaction has been found to offer a quite useful method for the catalytic asymmetric syntheses of polycyclic compounds. It is noteworthy that, very recently, the strategy described above has been successfully applied to an intermolecular version [13].

3 Catalytic Asymmetric Nitroaldol Reaction

A couple of years ago we started a research program concerning development of new reactions utilizing early transition metals. In particular, we were interested in the early transition metal alkoxide chemistry, which had not been widely studied. Early transition metal alkoxides are expected to show basicity, which will be useful in organic synthesis. When we started the research program, only Professor Stork had successfully utilized basicity of $Zr(O\text{-}n\text{-}Pr)_4$ for the synthesis of *trans*-hydrindans. $Zr(O\text{-}n\text{-}Pr)_4$ is known to exist as a trimer. On the other hand, $Zr(O\text{-}t\text{-}Bu)_4$, which we decided to study, exists as a monomer, and is expected to show stronger basicity than $Zr(O\text{-}n\text{-}Pr)_4$. In fact, in all the cases examined, reactions by $Zr(O\text{-}t\text{-}Bu)_4$ proceeded under milder reaction conditions than those by $Zr(O\text{-}n\text{-}Pr)_4$. Herein, three successful results utilizing $Zr(O\text{-}t\text{-}Bu)_4$ are described. Treatment of 2-cyclopentenone **50** (0.44 mmol) with 1.5 molar equiv of $Zr(O\text{-}t\text{-}Bu)_4$ in THF at -40 °C for 0.5 hr followed by stirring for 15 min with (*E*)-6-(benzyloxy)-2-hexen-1-al (**51**) (0.29 mmol) was found to afford the cross aldol adduct **52** in 64% yield (100% based on recovered **51**). Under these conditions, in contrast to the LDA assisted reaction (52% yield, 60% based on recovered **51**), neither self condensation products nor Michael products were obtained. The second example is the aldol reaction of the α-bromo ketone **54** with the aldehyde **53**. The introduction of a carbon-carbon triple bond at the C-13 position in prostacyclin derivatives is an important way to enhance their biological activity and metabolic stability. The hitherto known synthetic methods

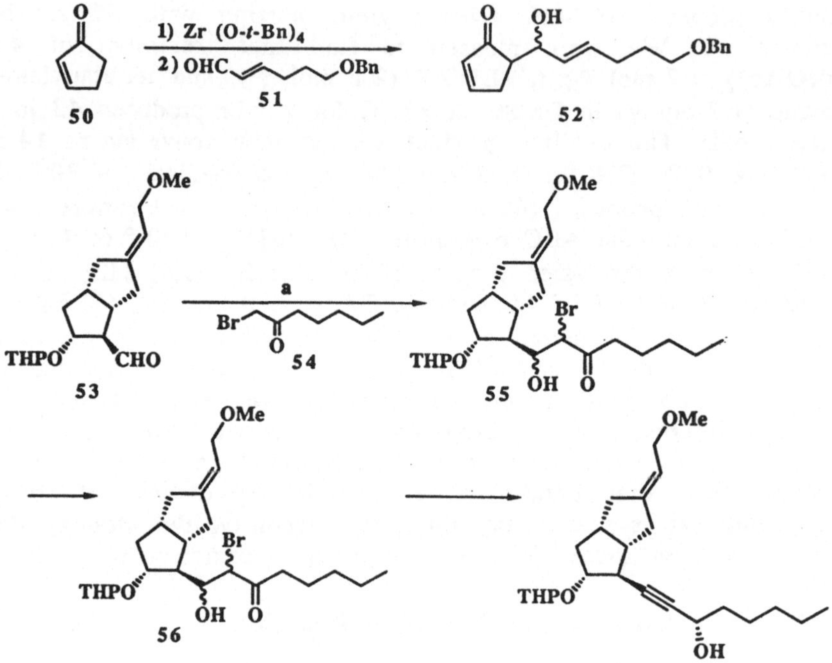

1) Zr (O-*t*-Bn)$_4$
2) OHC~~~OBn **51**

50 → **52**

53 → (a, **54**) → **55** → **56** →

a : Zr(O-*t*-Bu)$_4$ (2.5 molar equiv), THF, -30 °C, 30 min then 53, -30 °C, 20 min.

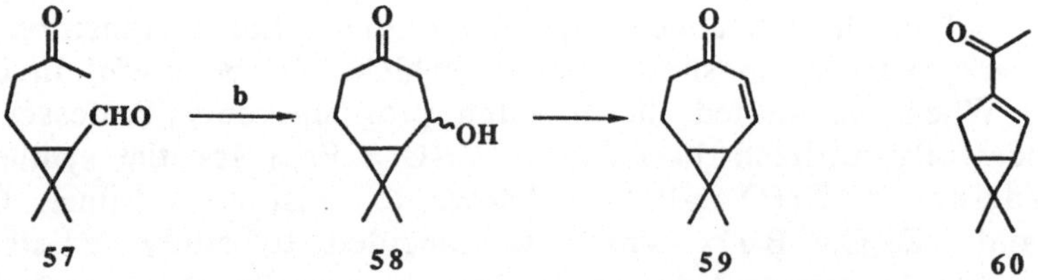

57 → (b) → **58** → **59**; **60**

b : Zr(O-*t*-Bu)$_4$ (2.0 molar equiv), THF, -30 °C — r.t., 6 h.

Scheme 12

starting with aldehydes involve dehydrobromination of the intermediate α-bromoenones. We planned to examine the production of α-bromo-enones by the Zr(O-*t*-Bu)$_4$-assisted cross aldol reaction. Treatment of the α-bromo ketone **54** (0.25 mmol) with Zr(O-*t*-Bu)$_4$ (0.25 mmol) in THF (-30 °C, 0.5 hr), followed by the addition of the aldehyde **53** (0.1 mmol) (-30 °C, 20 min), afforded a mixture of the α-bromo β-hydroxy ketone **55**, which was converted to the α-bromo enone **56** in 56% overall yield from **53**. Contrary to this results, attempts to construct the α-bromo enone **56** from the α-bromo ketone **54** and the aldehyde **53** using LDA, Sn(OTf)$_2$ or *n*-Bu$_2$BOTf were unsatisfactory. A further excellent example that shown the utility of $Zr(O\text{-}t\text{-}Bu)_4$ is a synthesis of (-)-8,8-dimethylbicyclo[5. 1. 0]oct-2-en-4-one (**59**), which is a useful intermediate for many natural products syntheses. The enone **59** has previously been synthesized from **57** by the Mukaiyama reaction in order to avoid the formation of the thermodynamically favorable five-membered-ring product **60**. We found that **59** could be readily prepared fron **57** via the intramolecular aldol adduct **58**, which was produced in 78% yield by treatment of **57** with 2 molar equiv of $Zr(O\text{-}t\text{-}Bu)_4$ in THF. Dehydration of **58** was carried out using NaOAc and Ac_2O in AcOH at 90 °C to give the desired enone **59** in 79% yield. Other metal enolates (Li, Cp_2ZrCl, ZnCl) generated by LDA or metal exchange from the corresponding lithium enolate gave **58** in less than 10% yield together with a mixture of the intermolecular aldol products and **57**. Another approach using $Sn(OTf)_2$ and $n\text{-}Bu_2BOTf$ gave **58** in trace and 13% yields, respectively [14]. Although $Zr(O\text{-}t\text{-}Bu)_4$ was found to be an efficient and convenient basic reagent in organic synthesis, all reactions examined were performed with stoichiometric quantities of the reagent. We envisioned that rare earth metal alkoxides would be stronger bases than group 4 metal alkoxides due to the lower ionization potential (ca. 5.4-6.4 eV) and the lower electronegativity (1.1-1.3) of rare earth metal elements; thus,

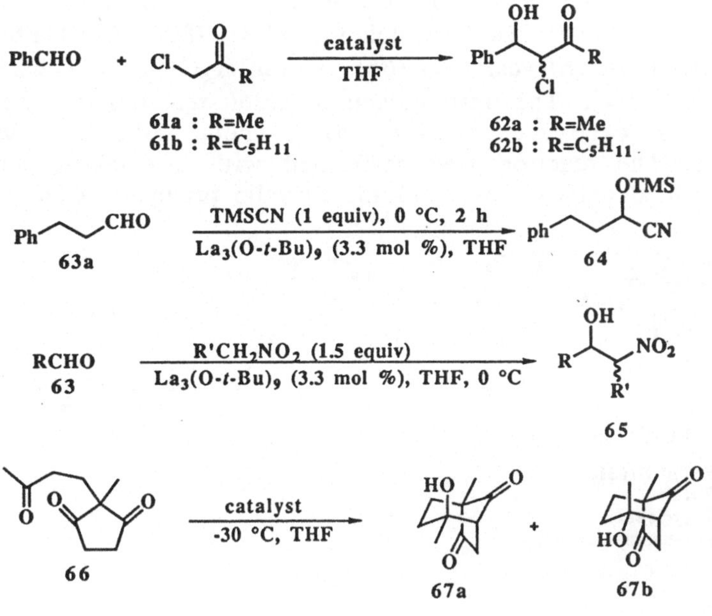

PhCHO + **61** —catalyst, THF→ **62**

61a : R=Me
61b : $R=C_5H_{11}$

62a : R=Me
62b : $R=C_5H_{11}$

63a —TMSCN (1 equiv), 0 °C, 2 h; $La_3(O\text{-}t\text{-}Bu)_9$ (3.3 mol %), THF→ **64**

RCHO **63** —$R'CH_2NO_2$ (1.5 equiv); $La_3(O\text{-}t\text{-}Bu)_9$ (3.3 mol %), THF, 0 °C→ **65**

66 —catalyst, -30 °C, THF→ **67a** + **67b**

Scheme 13

the catalytic use of rare earth metal alkoxides in organic synthesis was expected. Although a variety of rare earth metal alkoxides have been prepared for the last three decades [15], to our knowledge, there have been few reports concerning the basicity of rare earth metal alkoxides [16]. Herein, we describe several carbon-carbon bond-forming reactions catalyzed by rare earth metal alkoxides and their application to a catalytic asymmetric nitroaldol reaction. We began examination of the basicity of rare earth alkoxides by comparison with the basicity of $Zr(O\text{-}t\text{-}Bu)_4$. The intermolecular aldol reaction of the α-chloro ketones **61a** and **61b** with benzaldehyde was found to proceed smoothly in spite of the use of a catalytic amount of $La_3(O\text{-}t\text{-}Bu)_9$ or $Y_3(O\text{-}t\text{-}Bu)_8Cl$, giving the α-chloro-β-hydroxyketones **62a** : $La_3(O\text{-}t\text{-}Bu)_9$ (3.3 mol %), -72 °C, 30 hr, 74%; $Y_3(O\text{-}t\text{-}Bu)_8Cl$ (3.3 mol %), -43 °C, 3.5 hr, 50%; and **62b** : $La_3(O\text{-}t\text{-}Bu)_9$ (3.3 mol %), -50 °C, 4 hr, 83%. On the other hand, $Zr(O\text{-}t\text{-}Bu)_4$ exhibited much lower catalytic activity to give **62a** : 140 mol %, -50 °C, 2.5 hr, 86%; 10 mol %, -50 °C, 20 hr, 8%. Thus, the basicity of the rare earth alkoxides appears to be stronger than that of $Zr(O\text{-}t\text{-}Bu)_4$. Use of $Ti(O\text{-}t\text{-}Bu)_4$ afforded none of the coupling product. Likewise, the aldehyde **63a** was converted to the α-(trimethylsilyl)oxy nitrile **64** in 96% yield using a catalytic amount of $La_3(O\text{-}t\text{-}Bu)_9$. Furthermore, the various nitroaldol reactions proceeded smoothly by the use of a catalytic amount of $La_3(O\text{-}t\text{-}Bu)_9$, giving the following couplng products : **65a** R=$PhCH_2CH_2$, R'=H, 85%; **65b** R=Ph, R'=H, 68%; **65c** R=*i*-Pr, R'=H, 76%; **65d** R=cyclohexyl, R'=H, 78%; **65e** R=$PhCH_2CH_2$, R'=Me, 72% (*syn*:*anti*=ca. 1:1); **65f** R=$PhCH_2CH_2$, R'=$PhCH_2$, 76% (*syn*:*anti*=ca. 1:1). The intramolecular aldol reaction of the *meso*-triketone **66** catalyzed by rare earth metal alkoxides was also investigated. The reaction was performed with a catalytic amount of the rare earth alkoxides to afford the bicyclic products **67a** and **67b**

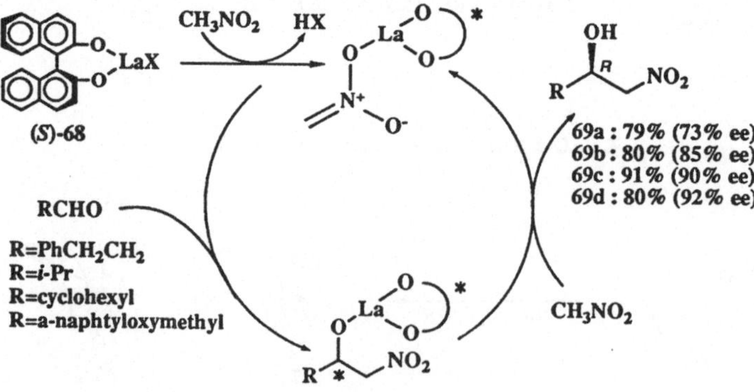

Scheme 14

$La_3(O\text{-}t\text{-}Bu)_9$ (3.3 mol %), 1 hr, **67a** (40%) and **67b** (30 %); $Y_3(O\text{-}t\text{-}Bu)_8Cl$ (3.3 mol %), 5 hr, **67a** (~100%); $Y_5(O\text{-}i\text{-}Pr)_{13}O$ (2 mol %), 4 hr, **67a** (94%); $Zr(O\text{-}t\text{-}Bu)_4$ (10 mol %), 24 hr, **67a** (30%). Use of $Al(O\text{-}t\text{-}Bu)_3$ gave none of the cyclized product. It is interesting to note that, in this case, the yttrium alkoxide was more effective than $La_3(O\text{-}t\text{-}Bu)_9$. Having established the several carbon-carbon bond-forming reactions catalyzed by the rare earth alkoxides, we turned our attention to a catalytic asymmetric synthesis using an optically active rare earth alkoxide. The optically active rare earth alkoxides **68** was prepared as follows. To a stirred suspension of $LaCl_3$ (3 mmol) in THF (10 ml) was added a solution of dilithium (*S*)-(-)-binaphthoxide (3 mmol) and NaO-*t*-Bu (3 mmol) in THF (30 ml). After addition of H_2O (30 mmol), the reaction mixture was stirred for 24 hr at room temperature, and the supernatant was used as a catalyst. The alkoxide **68** may exist as a trimer. However, the exact structure, whose ^{13}C-NMR spectrum was similar to that of the titanium binaphthoxide [17], has not been fully elucidated yet. Reactin of nitromethane (10 equiv) with cyclohexanecarboxaldehyde in the presence of **68** (10 mol %) was found to provide the (*R*) adduct **69c** with 90% ee in 91% yield (THF, -42 °C, 18 hr). Use of 2-methylpropanal afforded the (*R*) adduct **69b** with 85% ee in 80% yield, and use of hydrocinnamaldehyde gave the (*R*) adduct **69a** with 73% ee in 79% yield. Furthermore, α-naphtyloxyacetaldehyde was converted to the adduct **69d** with 92% ee in 80% yield, a potential intermediate for propranolol. To our best knowledge, this is the first example of a catalytic asymmetric nitroaldol reaction. The possible reaction mechanism is shown in Scheme 14, suggesting that the first step of the reaction would be ligand exchange [18]. We believe that this type of reaction will become a powerful synthetic methodology in organic synthesis as well as a novel lead for the catalytic asymmetric construction of C-C bonds. Many applications of this new catalytic methodology in synthesis are under investigation.

Acknowledgement

It is my pleasure to acknowledge the contributions of my many associates over the years, particularly Hiroaki Sasai, Yoshihiro Sato, Katsuji Kagechika, Mikiko Sodeoka, Takeyuki Suzuki.

References

[1] T. Katsuki, K. B. Sharpless, J. Am. Chem. Soc. 102 (1980) 5974
[2] R. Noyori, H. Takaya, Acc. Chem. Res. 23 (1990) 345, and references cited therein.
[3] K. Narasaka, Synthesis (1991) 1, and references cited therein.
[4] R. F. Heck, J. P. Nolly. Jr., J. Org. Chem. 37 (1972) 2320
[5] K. Karabelas, C. Westerlund, A. Hallberg, J. Org. Chem. 50 (1985) 1433; M. M.

Abelman, T. Oh, L. E. Overman, J. Org. Chem. 52 (1987) 4130
[6] Y. Sato, M. Sodeoka, M. Shibasaki, J. Org. Chem. 54 (1989) 4738
[7] Y. Sato, M. Sodeoka, M. Shibasaki, Chem. Lett. (1990) 1953
[8] G. A. Lawrence, Chem. Rev. 86 (1986) 17
[9] Y. Sato, S. Watanabe, M. Shibasaki, Tetrahedron Lett. 33 (1992) 2589
[10] Y. Sato, T. Honda, M. Shibasaki, Tetrahedron Lett. 33 (1992) 2593
[11] M. Shibasaki, T. Mase, S. Ikegami, J. Am. Chem. Soc. 108 (1986) 2090; T. Mase, M. Shibasaki, Tetrahedron Lett. 27 (1986) 5245
[12] K. Kagechika, M. Shibasaki, J. Org. Chem. 56 (1991) 4093
[13] F. Ozawa, A. Kubo, T. Hayashi, J. Am. Chem. Soc. 113 (1991) 1417; T. Hayashi, A. Kubo, F. Ozawa, Pure Appl. Chem. 64 (1992) 421; F. Ozawa, A. Kubo, T. Hayashi, Tetrahedron Lett. 33 (1992) 1485
[14] H. Sasai, Y. Kirio, M. Shibasaki, J. Org. Chem. 55 (1990) 5306
[15] R. C. Mehrotra, A. Singh, U. M. Tripathi, Chem. Rev. 91 (1991) 1287, and references cited therein.
[16] J. L. Namy, J. Souppe, J. Collin, H. B. Kagan, J. Org. Chem. 49 (1984) 2045
[17] K. Mikami, M. Terada, T. Nakai, J. Am. Chem. Soc. 112 (1990) 3949
[18] H. Sasai, T. Suzuki, S. Arai, T. Arai, M. Shibasaki, J. Am. Chem. Soc. 114 (1992) 4418

Application of High Pressure Techniques in Mechanistic and Synthetic Studies of Organometallic Systems in Solution

Rudi van Eldik

Institute for Inorganic Chemistry, University of Witten/Herdecke, Stockumer Str. 10, 5810 Witten, Germany

Summary

An account of our recent work dealing with the application of high pressure techniques in mechanistic studies of a wide range of organometallic reactions is presented. These include thermal ligand substitution, photo-induced ligand substitution, metal-carbon bond formation, and addition and elimination reactions. The construction of reaction volume profiles for some of the investigated reactions is reported, and the mechanistic meaning of such profiles is discussed. The characteristic pressure dependence of these reactions create the possibility to selectively tune the reactivity of, and product distribution for such organometallic systems. The effects are in many cases so significant that moderate pressures (up to 100 MPa) are sufficient to induce certain synthetic processes.

1 Introduction

Our group has over the past decade been intensively involved with the application of high pressure kinetic techniques in the study of inorganic, organometallic and bioinorganic reaction mechanisms in solution [1-5]. In general these studies have added a further dimension to mechanistic investigations by introducing pressure as a decisive parameter for the elucidation of the underlying reaction mechanisms. In addition, an improved understanding of the chemical process can fruitfully assist the selection of experimental conditions, the optimization of reaction product distribution, and synthesis of new materials. It is therefore the objective of this contribution to focus on the application of high pressure techniques in mechanistic studies and, as a result of that, in synthetic studies of organometallic systems in solution.

Much of the achieved advances in this area result from the development of instrumentation to study slow and fast reactions, including flow systems and relaxation techniques, at pressures up to 300 MPa. The techniques involve stopped-flow, T-jump, P-jump, NMR, ESR, flash-photolysis and pulse-radiolysis instrumentation [1, 6-9]. The fundamental principles involved in the application of pressure as a kinetic parameter, the determination of activation and reaction volumes, the construction of reaction volume profiles, and the interpretation of such data have been treated in detail elsewhere [1-4,6]. The basic idea is that the volume of activation represents the change in partial molar volume on going from the reactant to the transition state of the process, such that the reaction volume profile can be analysed in terms of volume changes along the reaction coordinate [3,4]. These can result from intrinsic volume changes due to changes in bond lengths and angles, and from solvational volume changes due to changes in solvent electrostriction as a result of the creation or neutralization of charges and dipole moments. These effects are observed experimentally as an increase or decrease in the observed rate or equilibrium constant as a function of pressure [3-6].

The examples discussed in this contribution will demonstrate the details of the involved principles. For this reason, a series of different types of reactions of organometallic systems that are induced thermally, photochemically or by pulse-radiolysis, has been selected for this presentation. It will become clear from these examples how the application of pressure can, on the one hand, improve our understanding of the underlying reaction mechanisms, and on the other hand, help us to tune the reactivity of a process to optimize product formation and distribution.

2 Ligand Substitution Reactions

Ligand substitution reactions of organometallic compounds have been the topic of many mechanistic investigations because of the fundamental importance of such reactions in many chemical and biological processes. For a general ligand substitution reaction, there are basically three simple pathways: (i) the dissociative (D) process with an intermediate of lower coordination number; (ii) the associative (A) process, with an intermediate of higher coordination numer; (iii) the interchange (I) process, in which no intermediate of lower or higher coordination number is involved, and in which either bond breakage (I_d) or bond formation (I_a) is the more important process. Such bond formation/bond breakage processes should be characterized by specific intrinsic volume changes, thus also by specific dependences on pressure.

Substitution reactions of a number of organometallic complexes and clusters have been investigated using high pressure kinetic techniques. For instance, the substitution of coordinated CO by $P(OMe)_3$ on $Cr(CO)_4$phen is characterized by a volume of activation ($\Delta V^{\#}$) of $+13.8 \pm 0.5$ cm^3mol^{-1}, whereas the reverse reaction exhibits a $\Delta V^{\#}$ of $+19.2 \pm 0.5$ cm^3mol^{-1} [10]. A volume profile for this system (Figure 1) clearly demonstrates the significantly higher partial molar volume of the transition state as compared to either the reactant or product states and the operation of a dissociative (D) substitution mechanism.

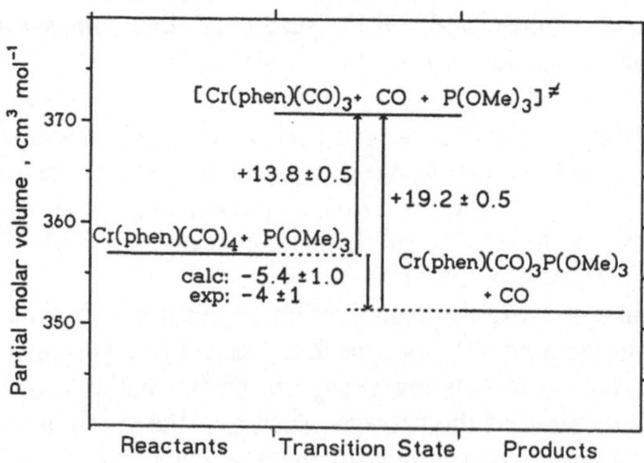

Figure 1. Volume profile for the reaction [10]
$Cr(phen)(CO)_4 + P(OMe)_3 \rightarrow Cr(phen)(CO)_3P(OMe)_3 + CO$

The overall reaction volume calculated from the difference in $\Delta V^{\#}$ for the forward and reverse reactions, or from the difference in partial molar volumes of reactants and products, is slightly negative and indicates an overall volume decrease during the substitution process.

Earlier work on the substitution of CO on triruthenium clusters [11] also resulted in large positive $\Delta V^{\#}$ values for rate-determining CO dissociation, viz. $HRu_3(CO)_{11}^-$ (+21.2 ± 1.4), $Ru_3(CO)_{11}(CO_2CH_3)^-$ (+16 ± 2) and $Ru_3(CO)_{10}(P(OMe)_3)(CO_2CH_3)^-$ (+24 ± 2 cm^3mol^{-1}).

On the contrary, a number of typical associative (A) substitution reactions are characterized by significantly negative volumes of activation, which means that such reactions are significantly accelerated by pressure. For instance, typical $\Delta V^{\#}$ values between -14 and -17 cm^3mol^{-1} were found for ligand substitution on $Rh^I(Cp)(CO)_2$, which is generally accepted to proceed according to an associative ring-slippage mechanism [12]. More recently we investigated a series of associative substitution reactions on an osmium cluster of the type in (1), which resulted in $\Delta V^{\#}$ values between -8 and -32 cm^3mol^{-1}, depending on the

$$(\mu\text{-H})_2Os_3(CO)_{10} + L \rightarrow (\mu\text{-H})HOs_3(CO)_{10}L \qquad (1)$$

nucleophilicity and especially the cone angle of L [13]. Once a feeling for the meaning of $\Delta V^{\#}$ in terms of well-understood ligand substitution processes has been developed, it is possible to apply the technique to systems for which less mechanistic information is available.

Our interest in the mechanistic behaviour of solvento metal carbonyl complexes led to a series of high pressure studies involving the replacement of coordinated solvent molecules. In most cases such species are short-lived and can only be studied by flash-photolysis (see next section). However, in the case of THF as solvent, complexes of the type $M(CO)_5THF$ are fairly stable, can be prepared in solution, and their substitution behaviour can be studied using stopped-flow techniques [14]. A systematic variation of M(Cr, Mo, W) and the entering nucleophile resulted in the data summarized in Table 1, from which some interesting trends can be observed. The $\Delta V^{\#}$ data show a general trend to more negative values in going to the larger metal center (Cr < Mo < W) and larger entering nucleophiles. Thus bond formation seems to be more predominant for these systems, and a gradual

Table 1. Summary of rate and activation parameters for the reaction [14]

$$M(CO)_5THF + L \xrightarrow{k} M(CO)_5L + THF$$

M	Ligand	k at 25 °C $M^{-1}sec^{-1}$	$\Delta H^{\#}$ kJ mol^{-1}	$\Delta S^{\#}$ J $K^{-1}mol^{-1}$	$\Delta V^{\#}$ cm^3mol^{-1}
Cr	piperidine	1.53	45 ± 2	-90 ± 8	-2.2 ± 0.6
	$P(C_6H_5)_3$	0.36	78 ± 1	+8 ± 8	-1.9 ± 1.0
	$P(OC_2H_5)_3$	0.37	53 ± 1	-74 ± 2	-3.6 ± 0.7
Mo	piperidine	15.2	60 ± 2	-21 ± 8	-3.6 ± 1.2
	$P(C_6H_5)_3$	2.5	65 ± 4	-17 ± 14	-8.3 ± 1.0
	$P(OC_2H_5)_3$	1.7	55 ± 2	-55 ± 8	-5.8 ± 0.9
W	piperidine	0.75	38 ± 6	-122 ± 20	-4.4 ± 0.5
	$P(C_6H_5)_3$	0.071	50 ± 3	-100 ± 11	-12.2 ± 0.4
	$P(OC_2H_5)_3$	0.31	49 ± 3	-89 ± 10	-14.9 ± 1.0

changeover to a more associative substitution mechanism from Cr to W was suggested. An appropriate mechanism is outlined in (2), for which the rate law is given in (3). The results in Table 1 furthermore demonstrate that $\Delta S^{\#}$ does not exhibit such a clear trend as in the case of $\Delta V^{\#}$, an observation that has been made in many mechanistic studies [3-6].

$$\begin{array}{ccc} M(CO)_5THF & \underset{k_{-1}}{\overset{k_1}{\rightleftharpoons}} & M(CO)_5 + THF \\ {\scriptstyle +L,\ -THF}\ \downarrow k_3 & & \downarrow k_2\ {\scriptstyle +L} \\ & M(CO)_5L & \end{array} \quad (2)$$

$$k_{obs} = \{ k_1k_2 /k_{-1} + k_3\} [L] \quad (3)$$

3 Photo-Induced Ligand Substitution Reactions

Flash photolysis techniques have in general been adopted with great success to study the substitution behaviour of reactive intermediates in organometallic systems. Irradiation of $M(CO)_6$ (M = Cr, Mo, W) in a coordinating solvent S produces intermediates of the type $M(CO)_5S$ [15], which can undergo rapid solvent displacement with L to produce $M(CO)_5L$. The effect of pressure on such substitution reactions has been studied for a series of M, S and L (Table 2), and the results clearly demonstrate the crucial role played by the size of the metal center M, the bulkiness of L and the binding properties of S [16, 17]. Once again the data in Table 2 exhibit the same trend as in Table 1, viz. to more negative (less positive) values for the larger metal centers.

Table 2. Summary of $\Delta V^{\#}$ data for solvent displacement reactions of the type [16, 17] $M(CO)_5S + L \rightarrow M(CO)_5L + S$

S	L	$\Delta V^{\#}$, cm^3mol^{-1}		
		M = Cr	M = Mo	M = W
fluorobenzene	1-hexene	+9.4 ± 0.7	+5.8 ± 0.8	+2.5 ± 0.2
toluene		+10.8 ± 0.7	+3.2 ± 0.3	
benzene		+10.9 ± 1.0		
chlorobenzene		+5.4 ± 0.4	+3.2 ± 0.3	+0.4 ± 0.3
n-heptane		+6.2 ± 0.2	+2.2 ± 0.3	+2.7 ± 0.4
fluorobenzene	piperidine	+6.1 ± 0.3		
toluene		+4.8 ± 1.4		
benzene		+4.2 ± 0.3		
chlorobenzene		+0.2 ± 0.2		
n-heptane		+1.4 ± 0.4		

It is also possible to displace the coordinated solvent molecule via ring-closure of a potential bidentate ligand such as a P-olefin. A series of such reactions for cis-$(CO)_4W(S)(PPh_2(CH_2)_nCH = CH_2)$ (n = 1 to 4, S = chlorobenzene) all significantly slow down on increasing pressure, with corresponding $\Delta V^{\#}$ values of +7.7 (n = 1), +5.1 (n = 2), +10.7 (n = 3) and +10.5 (n = 4) cm^3mol^{-1} [18]. These results indicate that chelate ring-closure for n = 1, 2 mainly involves an interchange (I_d) mechanism in which the olefin moiety is pre-associated with the metal center followed by rate-determining loss of chlorobenzene. In the case of ring-closure for n = 3, 4, $\Delta V^{\#}$ reaches the limiting value observed for the dissociation of chlorobenzene and therefore presumably does not involve any significant pre-association [18].

When the attacking nucleophile is a bidentate ligand, flash photolysis of $M(CO)_6$ results in the reaction sequence outlined in (4), in which ring-closure now involves CO displacement as compared to solvent displacement discussed above. Typical data for a series of ring-

$$\begin{aligned} M(CO)_6 &\xrightarrow{h\nu} M(CO)_5 + CO \\ M(CO)_5 + S &\xrightarrow{fast} M(CO)_5\,S \\ M(CO)_5\,S + N\text{-}N &\xrightarrow{k_1} M(CO)_5N\text{-}N + S \\ M(CO)_5N\text{-}N &\xrightarrow{k_2} M(CO)_4(N\text{-}N) + CO \end{aligned} \qquad (4)$$

closure reactions in Table 3, indicate that the nature of the metal center and the bulkiness of N-N control the intimate nature of the CO displacement mechanism [19,20]. The larger metal centers (Mo and W) tend to ring-close in an associative way and exhibit significantly negative $\Delta V^{\#}$ values. The smaller Cr center must loose CO prior to ring-closure, unless there is no bulkiness on the entering ligand, as in the case of ethylenediamine, that prevents an associative ring-closure reaction. More recently, a systematic study of the effect of bulky substituents on the bpy and phen ligands for the ring-closure reaction of $Mo(CO)_5N$-N was undertaken [21,22]. The results indicate a significant effect in the case of bypiridine, but almost no effect in the case of phenanthroline.

Table 3. Summary of $\Delta V^{\#}$ data for the ring-closure step in reaction (4) [19,20]

M	Solvent	N-N [a]	k_2 at 25 °C s^{-1}	$\Delta H^{\#}$ kJ mol^{-1}	$\Delta S^{\#}$ J $K^{-1}mol^{-1}$	$\Delta V^{\#}$ cm^3mol^{-1}
Cr	toluene	en	1.6×10^{-5}	57 ± 5	-145 ± 17	-11.9 ± 1.5
	toluene	$dabR_2$	1.5×10^{-4}	81 ± 2	-47 ± 7	+17.2 ± 1.0
	fluorobenzene	phen	26	52 ± 2	-42 ± 6	+6.2 ± 0.5
Mo	toluene	en	3.0×10^{-5}	72 ± 7	-92 ± 22	-5.4 ± 0.8
	toluene	$dabR_2$	1.1×10^{-3}	78 ± 5	-40 ± 17	-9.5 ± 0.4
	fluorobenzene	phen	1.1×10^{4}	47 ± 2	-9 ± 7	-2.9 ± 0.2
W	toluene	en	3.0×10^{-6}	54 ± 2	-172 ± 16	-12.3 ± 1.4
	toluene	$dabR_2$	2.9×10^{-5}	83 ± 8	-53 ± 26	-13.7 ± 1.3
	toluene	bpy	8.0×10^{-2}	58 ± 2	-74 ± 7	-10.9 ± 1.1
	fluorobenzene	phen	4.3×10^{2}	51 ± 2	-23 ± 5	-8.2 ± 0.2

[a]Abbreviations: en = ethylenediamine, $dabR_2$ = 1,4-diisopropyl-1,4-diazabutadiene, bpy = bipyridine, phen = 1,10-phenanthroline

Ligand substitution reactions that occur in the electronic excited state of organometallic complexes also exhibit very characteristic pressure dependences [23]. In order to demonstrate this, the photosubstitution of CO in $M(CO)_4phen$ (M = Cr, Mo, W) was investigated, since the photoactivity of the lower lying MLCT states has been a controversial issue in the literature [24]. On the one hand, it can be assumed that excitation of the MLCT state is followed by thermal back population of the higher energy LF state from which a dissociative photosubstitution reaction occurs. On the other hand, it was argued that the MLCT states themselves are photoactive and could undergo substitution in an associative way. The pressure dependence of the photosubstitution quantum yield was measured as a function of irradiation wavelength in order to populate the different excited states. Indeed, the quantum yields measured for LF and MLCT excitation revealed very different pressure dependencies, and the corresponding volumes of activation are summarized in Table 4. The

Table 4. Quantum yields and volumes of activation for the reaction in toluene [24]

$$M(CO)_4phen + PEt_3 \xrightarrow{h\nu} M(CO)_3(PEt_3)phen + CO$$

M	λ_{irr} nm	$\Delta V^{\#}(MLCT)$	$\Delta V^{\#}(LF)$ cm^3mol^{-1}
Cr	366		+9.6 ± 1.6
	546	+2.7 ± 0.3	
Mo	366		+6.0 ± 0.2
	546	-13.3 ± 1.2	
W	366		+8.2 ± 0.5
	546	-12.0 ± 0.7	

results clearly demonstrate that in the Mo and W complexes MLCT and LF photosubstitution occur according to associative and dissociative mechanisms, respectively. For the smaller Cr complex, the associative MLCT path does not seem to be possible and even this reaction has to follow a dissociatively activated process, presumably of the interchange (I_d) type. These results nicely demonstrate the value of pressure as a key kinetic parameter to distinguish between associative and dissociative photosubstitution mechanisms.

4 Metal-Carbon Bond Formation Reactions

The mechanisms of inorganic and organometallic free radical reactions have in many cases been studied using pulse-radiolysis techniques, but have not been fully elucidated. A series of investigations were therefore undertaken to use pressure as a mechanistic indicator in such reactions. Relevant for this presentation are reactions that involve the formation and breakage of metal-carbon σ bonds. For instance, for the overall reaction (5), the reaction

$$Ni^{II}(cyclam)^{2+} + \cdot CH_3 + H_2O \rightleftharpoons Ni^{III}(cyclam)(CH_3)H_2O^{2+} \quad (5)$$

volume was found to be -19.7 cm^3mol^{-1}, whereas $\Delta V^{\#}$ for the forward and reverse reactions equals +4.7 and +24.4 cm^3mol^{-1}, respectively [25]. These values indicate that metal-carbon bond formation is significantly assisted by pressure since the reverse homolysis reaction is inhibited by pressure. Similar results were found for the reaction of $Co^{II}(nta)(H_2O)_2^-$ with $\cdot CH_3$ to produce $Co^{III}(nta)(H_2O)CH_3^-$ [26]. The volume profile for this reaction is presented in Figure 2, which clearly indicates the significantly higher partial molar volume of the transition state. It follows that the displacement of the coordinated solvent molecule on the Co(II) complex presumably occurs according to an I_d mechanism. The large volume increase during the reverse homolysis reaction is partly ascribed to desolvation of the aliphatic free radical and partly to Co(III) expanding to Co(II).

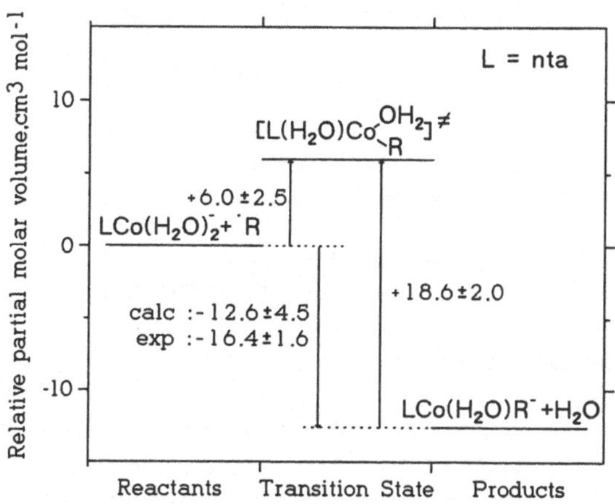

Figure 2. Volume profile for the reaction [26]
$Co(nta)(H_2O)_2^- + \cdot CH_3 \rightarrow Co(nta)(H_2O)CH_3^- + H_2O$

Recently, a systematic study of the effect of pressure on the formation of organochromium(III) species was undertaken [27]. This reaction was studied for 10 different aliphatic radicals $\cdot R$ that react with $Cr(H_2O)_6^{2+}$ to form $Cr(H_2O)_5R^{2+}$. The reactions all exhibited a small positive volume of activation with an average value of $+4.3 \pm 1.0$ cm^3mol^{-1}, independent of the nature of R. This constant value and the fact that the bond formation rate constants are all very similar and close to that for water exchange on $Cr(H_2O)_6^{2+}$, suggested that the metal-carbon bond formation is controlled by solvent exchange on $Cr(H_2O)_6^{2+}$, which follows an I_d mechanism. This process is presumably enhanced by Jahn-Teller distortion in $Cr(H_2O)_6^{2+}$. Combining these results with that of an earlier homolysis study, resulted in the volume profile reported in Figure 3. The latter clearly demonstrates the dissociative nature of the complex formation process. In addition, the large volume collapse on the formation of the metal-carbon bond must be due to the intrinsic volume decrease due to Cr-R bond formation and the contraction of the chromium metal due to $Cr^{II}\text{-}R \rightarrow Cr^{III}\text{-}R^-$ [27].

Finally, the pressure dependence of the spontaneous and acetate-catalyzed heterolysis reaction of $Cr\text{-}R^{2+}$ was also investigated [28]. The reported $\Delta V^{\#}$ data clearly indicate that coordinated acetate causes a trans-labilization effect in which coordinated R is substituted by water in a dissociative way.

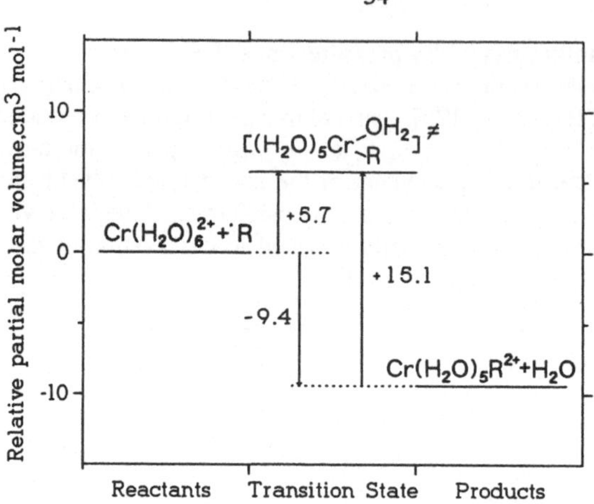

Figure 3. Volume profile for the reaction [27]
$Cr(H_2O)_6^{2+}$ + $\cdot C(CH_3)_2OH \rightarrow Cr(H_2O)_5C(CH_3)_2OH^{2+}$ + H_2O

5 Addition and Elimination Reactions

The above outlined examples have clearly demonstrated that bond formation processes are characterized by negative volumes of activation, whereas bond breakage processes are characterized by positive values. It is therefore not surprising that oxidative addition reactions are characterized by significantly negative volumes of activation, partly due to bond formation and partly due to a decrease in the ionic radius of the metal center during the oxidation process [1,3]. Typical volumes of activation can be as large as -30 to -40 cm^3mol^{-1}, and will in general depend on the nature of the solvent due to charge creation in the transition state [29]. Similarly, cyclo-addition reactions of for instance $Fe(CO)_3$HTE (HTE = heptatrienone) with tetracyanoethane are characterized by volumes of activation between -29 and -33 cm^3mol^{-1} [30], once again emphasizing the importance of bond formation in the transition state.

A recent study of [2+2] cyclo-addition reactions (6) on the coordinated ligand of pentacarbonyl carbene complexes of Cr and W, revealed a significant acceleration by pressure and almost no dependence on the polarity of the solvent [31]. The average $\Delta V^{\#}$ value of -16 ± 1 cm^3mol^{-1} and the solvent independence of the processes suggested that the

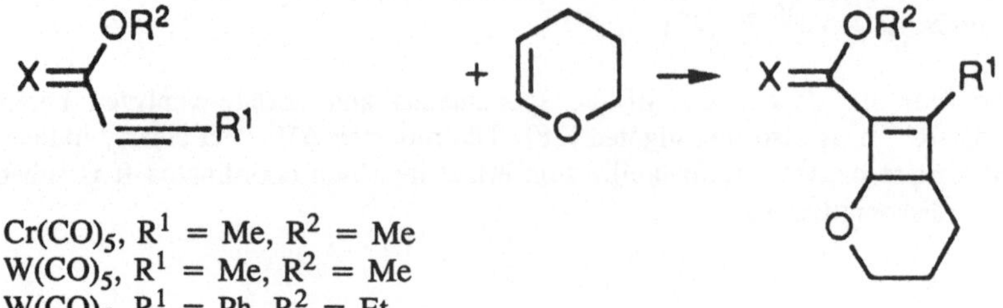

(6)

$X = Cr(CO)_5$, R^1 = Me, R^2 = Me
$X = W(CO)_5$, R^1 = Me, R^2 = Me
$X = W(CO)_5$, R^1 = Ph, R^2 = Et

reaction follows a nonpolar concerted, synchronous one-step mechanism. The observed pressure acceleration is very similar to that observed for the insertion of dipropylcyanamide and 1-(diethylamino)propyne into the metal-carbene bond of pentacarbonyl(methoxyphenylcarbene)chromium and -tungsten for which $\Delta V^{\#}$ varies between -17 and -25 cm^3mol^{-1} [32].

We recently performed a detailed study of the effect of pressure on addition reactions of α,β-unsaturated Fischer carbene complexes [33]. The reactions in (7) when performed in

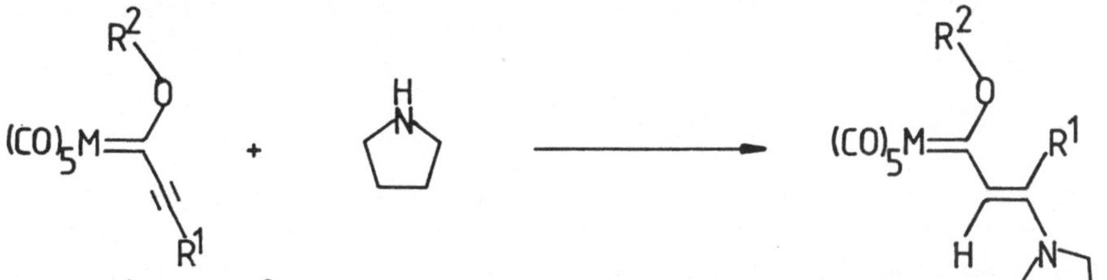

(7)

M	R^1	R^2
Cr	Ph	Et
Mo	Ph	Et
W	Ph	Et
W	Me	Me

acetonitrile all exhibit $\Delta V^{\#}$ values between -15 and -17 cm^3mol^{-1}, and are characterized by extremely low $\Delta H^{\#}$ (between 2 and 10 kJ mol^{-1}) and very negative $\Delta S^{\#}$ values (between -121 and -152 J $K^{-1}mol^{-1}$). In addition, a systematic solvent dependence study was performed for one of the systems, and the reaction showed a 50 fold decrease in going from acetonitrile to n-heptane, accompanied by a significant decrease in $\Delta V^{\#}$ (to more negative values). The latter data exhibit a good correlation with the solvent parameter q_p (see Figure 4), from which it follows that ΔV_{intr} is approx. -14 cm^3mol^{-1}. These observations and the absence of a meaningful kinetic isotope effect when deuterated pyrrolidine is used, indicate that the reaction follows a two step process with a polar transition state leading to a zwitterionic intermediate.

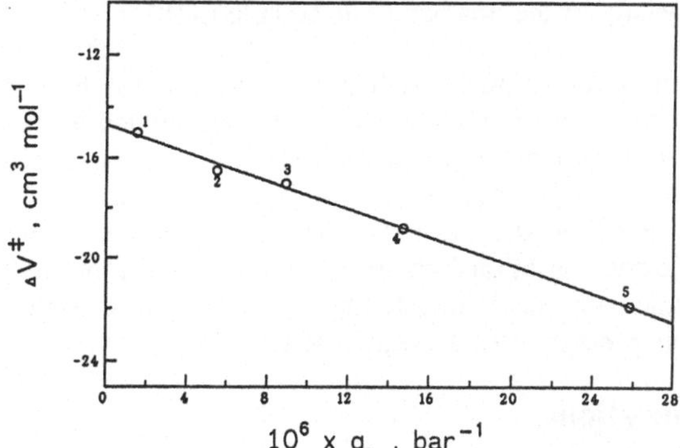

Figure 4. Plot of $\Delta V^{\#}$ versus q_p for reaction (7) with M = W, R^1 = Ph and R^2 = Et [33] Solvents: acetonitrile (1), 1,2-dichlorobenzene (2), chlorobenzene (3), benzene (4), n-heptane (5)

A similar study for the addition of a series of p-substituted anilines to a Fischer carbene complex as shown in (8), resulted in $\Delta V^{\#}$ values between -21 and -27 cm^3mol^{-1} [33], i.e.

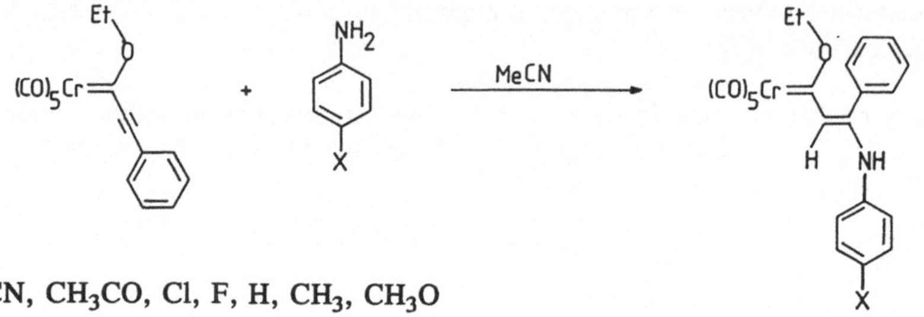

(8)

X = CN, CH_3CO, Cl, F, H, CH_3, CH_3O

significantly more negative than that reported above for the corresponding reaction with pyrrolidine. However, the second-order rate constants exhibit an excellent correlation with the basicity of the amine (expressed as pK_a value in H_2O), which is accompanied by a decrease in $\Delta H^{\#}$ and an increase in $\Delta V^{\#}$ (to more positive values) with increasing basicity. The trend in $\Delta V^{\#}$ can be correlated with an early or late transition state for the fast and slow addition reactions, respectively, which will result in more negative $\Delta V^{\#}$ values for the slower (later transition state) reactions [33]. These trends and the absence of a significant kinetic isotope effect suggest that a similar mechanism as suggested for the addition of pyrrolidine is operative.

The effect of pressure on the addition of a series of substituted pyridines to tricarbonyl(1-5-η-dienyl)iron(II) (dienyl = C_6H_7, 2-$MeOC_6H_6$ and C_7H_9) was investigated [34-37]. The reactions exhibited no significant pressure dependence for the addition and dissociation of 4-cyanopyridine [34], but revealed significantly negative and positive $\Delta V^{\#}$ values for these reactions with 4-formylpyridine, respectively [36]. In the case of 4-ethylpyridine the addition reactions exhibit a slightly negative volume of activation [35]. This work demonstrates that amine basicity controls the location of the transition state during the addition reactions. The mechanism can therefore vary from a pure interchange of bonds on the coordinated dienyl ligand (i.e. a zero $\Delta V^{\#}$) to a highly associative bond formation process (negative $\Delta V^{\#}$), depending on the basicity of the attacking nucleophile [37].

By way of comparison, elimination and reductive elimination reactions are expected to show significantly positive volumes of activation and to be decelerated by pressure. For instance, reductive elimination of H_2 from $H_3Ru_3(\mu_3\text{-}COMe)(CO)_9$ is characterized by a $\Delta V^{\#}$ value of +20 ± 2 cm^3mol^{-1}, whereas the reverse hydrogenation of $HRu_3(\mu\text{-}COMe)(CO)_{10}$ exhibits a $\Delta V^{\#}$ of +9.6 ± 0.6 cm^3mol^{-1}, which is consistent with a CO dissociation mechanism [38]. Hydride migration from bridging to terminal coordination modes can result in a significant volume increase as reflected by the recently reported $\Delta V^{\#}$ value of +4.1 ± 0.3 cm^3mol^{-1} for this process in $(\mu\text{-}H)_2Ru_3(\mu_3\text{-}CHCO_2Me)(CO)_9$ [39].

6 Synthetic Applications

The characteristic pressure dependence of the typical reactions discussed in the previous sections creates the possibility to employ the acceleration or deceleration by pressure in synthetic applications to tune the selectivity for a particular reaction product. This has been done for many organic systems [3,40] and to some extent in recent years for organometallic systems [41,42]. In general, a $\Delta V^{\#}$ value of -10 cm^3mol^{-1} corresponds to an increase in rate constant of a factor of two at a moderate pressure of 100 MPa (1 kbar). For instance, typical oxidative addition reactions exhibit $\Delta V^{\#}$ values as large as -40 cm^3mol^{-1}, which will cause a significant acceleration by pressure even at only a few hundred atmospheres.

The addition reactions to the coordinated ligand of metal carbene complexes discussed in the previous section, are all significantly accelerated by pressure. Various techniques can be employed to remove the metal carbonyl fragment from the organic addition product [43]. For instance hydrogenation with $NaBH_4$ leads to the formation of the corresponding ether, or treatment with HBr leads to the corresponding aldehyde. From a synthetic point of view, oxidative cleavage is the most convenient way to release the carbene moiety. Oxidizing agents include pyridine N-oxide, DMSO and Ce(IV) compounds, and lead to the formation of esters.

A series of organic synthetic reactions using organometallic reagents have been investigated under pressures as high as 2500 MPa (25 kbar) [42]. In this way aldol condensations can be performed under neutral conditions for molecules that are acid - or base - sensitive [44]. Similarly, high pressure techniques could be employed to prepare [4+2] cycloadducts via the Diels-Alder reaction between Danishefsky's diene and butyl glyoxylate as shown in (9) [45].

Me3SiO ... + H–C(=O)–CO2Bu^n ⟶ Me3SiO ... CO2Bu^n (OMe) + Me3SiO ... CO2Bu^n (OMe) (9)

The product distribution in the [4+2] cycloaddition of (-)-menthyl ester of (Z) -3-tributylstannylacrylic acid to 2,3-dimethyl-1,3-diene, leading to a mixture of diastereoisomers, depends on the applied pressure [46]. In the case of heat sensitive organotin compounds, good yields of cyclo-addition products, see reaction (10), could be

R1, R2, R3 diene + PhCH=CHSnBu3^n ⟶ R1, R2, R3 cyclohexene (Ph, SnBu3^n) (10)

R^1=H, R^2=R^3=Me	Z:E=70:30	0.1 MPa, 180 °C, 30 %	cis: trans = 19 : 81
	Z:E=80:20	2300 MPa, 70°C, 80 %	cis: trans = 90 : 10
R^1=OMe, R^2=R^3=H	Z:E=75:25	0.1 MPa, 180 °C, 42 %	cis: trans = 22 : 78
	Z:E=75:25	2000 MPa, 50 °C, 80 %	cis: trans = 70 : 30

obtained under high-pressure conditions, whereas reactions at ambient pressure and high temperature resulted in low yields and reverse diastereoselectivity due to isomerization of the starting dienophiles [42]. Application of pressure also had a significant affect on various substitution reactions, viz. the silylation of tertiary alcohols [47], and rearrangement processes, viz. the synthesis of α-silylated esters and lactones [48]. These examples demonstrate the versatility of the application of high pressure techniques in the synthesis of organic/organometallic materials.

7 Conclusions

The results reported in the previous sections for the different type of reactions of organometallic systems have clearly demonstrated that the additional physical parameter pressure can add a decisive dimension to mechanistic studies of such systems. The possibility to correlate the volume of activation with the partial molar volumes of the reactant and product species, or with the volume of activation for the reverse reaction, has led to the construction of reaction volume profiles, which create the possibility to visualize the chemical process in terms of volume changes along the reaction coordinate.

The fact that the rate-determining step of a particular process exhibits a characteristic pressure dependence, creates the possibility to tune the reactivity of particular systems via the application of moderate pressures. This can lead to the selective synthesis of particular reaction products in cases where the product distribution proves to be pressure dependent. Some of the examples presented here exhibit a large pressure sensitivity, meaning that even the application of a moderate pressure could be extremely appropriate in synthetic processes. In addition, $\Delta V^{\#}$ data and the revealed mechanistic information should be of predictive value in the optimization of the design of indrustrial chemical processes. In this way more moderate experimental conditions in terms of temperature can be reached, and products can be synthesized that are practically not possible at ambient pressure.

Acknowledgements

The author gratefully acknowledges the substantial financial support from the Volskwagen-Stiftung. Support from the Deutsche Forschungsgemeinschaft, Fonds der Chemischen Industrie, Max-Buchner Forschungsstiftung and the German-Israeli Foundation is also kindly acknowledged. The work discussed in this contribution was performed in collaboration with G.R. Dobson (Denton, USA), P.C. Ford (Santa Barbara, USA), J.B. Keister (Buffalo, USA), W.D. Wulff (Chicago, USA), A.J. Poë (Toronto, Canada), J.G. Leipoldt (Bloemfontein, South Africa), D. Meyerstein/H. Cohen (Beer-Sheva, Israel), and H. Fischer (Konstanz, Germany).

References

[1] R. van Eldik (Ed.): Inorganic High Pressure Chemistry. Kinetics and Mechanisms, Elsevier, Amsterdam (1986)
[2] R. van Eldik, J.J. Jonas (Eds.): High Pressure Chemistry and Biochemistry, Reidel, Dordrecht (1987)
[3] R. van Eldik, T. Asano, W.J. le Noble, Chem. Rev. 89 (1989) 549
[4] R. van Eldik, A.E. Merbach, Comments Inorg. Chem. 12 (1992) 341
[5] R. van Eldik, Perspectives in Coordination Chemistry, Verlag Chemie, Weinheim, in press
[6] J.W. Akitt, A.E. Merbach, NMR Basic Principles and Progress, Springer-Verlag, Berlin, Heidelberg, 24 (1990) 189
[7] M. Kotowski, R. van Eldik, Coord. Chem. Rev. 93 (1989) 19
[8] M. Spitzer, F. Gärtig, R. van Eldik, Rev. Sci. Instrum. 59 (1989) 2092
[9] J.F. Wishart, R. van Eldik, Rev. Sci. Instrum. 63 (1992) 3224
[10] K.J. Schneider, R. van Eldik, Organometallics 9 (1990) 1235
[11] D.J. Taube, R. van Eldik, P.C. Ford, Organometallics 6 (1987) 125
[12] P. Vest, J. Anhaus, H.C. Bajaj, R. van Eldik, Organometallics 10 (1991) 818
[13] A. Neubrand, R. van Eldik, A.J. Poë, unpublished results

[14] S. Wieland, R. van Eldik, Organometallics 10 (1991) 3110
[15] S. Wieland, R. van Eldik, J. Phys. Chem. 94 (1990) 5865
[16] S. Zhang, G.R. Dobson, H.C. Bajaj, V. Zang, R. van Eldik, Inorg. Chem. 29 (1990) 3477
[17] S. Zhang, V. Zang, H.C. Bajaj, G.R. Dobson, R. van Eldik, J. Organomet. Chem. 397 (1990) 279
[18] V. Zang, S. Zhang, C.B. Dobson, G.R. Dobson, R. van Eldik, Organometallics 11 (1992) 1154
[19] K. Bal Reddy, R. van Eldik, Organometallics 9 (1990) 1418
[20] S. Zhang, V. Zang, G.R. Dobson, R. van Eldik, Inorg. Chem. 30 (1991) 355
[21] K. Bal Reddy, R. Hoffmann, G. Konya, R. van Eldik, E.M. Eyring, Organometallics 11 (1992) 2319
[22] K. Bal Reddy, B. Ryan Brady, E.M. Eyring, R. van Eldik, J. Organomet. Chem. in press
[23] S. Wieland, R. van Eldik, Coord. Chem. Rev. 97 (1990) 155
[24] S. Wieland, K. Bal Reddy, R. van Eldik, Organometallics 9 (1990) 1802
[25] R. van Eldik, H. Cohen, A. Meshulam, D. Meyerstein, Inorg. Chem. 29 (1990) 4156
[26] R. van Eldik, H. Cohen, D. Meyerstein, Angew. Chem. Int. Ed. Engl. 30 (1991) 1158
[27] R. van Eldik, W. Gaede, H. Cohen, D. Meyerstein, Inorg. Chem. in press
[28] H. Cohen, W. Gaede, A. Gerhard, D. Meyerstein, R. van Eldik, Inorg. Chem. in press
[29] J.A. Venter, J.G. Leipoldt, R. van Eldik, Inorg. Chem. 30 (1991) 2207
[30] N. Hallinan, P. Mc Ardle, J. Burgess, P. Guardado, J. Organomet. Chem. 333 (1987) 77
[31] R. Pipoh, R. van Eldik, S.L.B. Wang, W.D. Wulff, Organometallics 11 (1992) 490
[32] K.J. Schneider, A. Neubrand, R. van Eldik, H. Fischer, Organometallics 11 (1992) 267
[33] R. Pipoh, R. van Eldik, prepared for publication
[34] T.I. Odiaka, R. van Eldik, J. Organomet. Chem. 425 (1992) 89
[35] T.I. Odiaka, R. van Eldik, J. Chem. Soc., Dalton Trans. in press
[36] T.I. Odiaka, R. van Eldik, J. Organomet. Chem. in press
[37] T.I. Odiaka, R. van Eldik, J. Chem. Soc., Dalton Trans. submitted for publication
[38] J. Anhaus, H.C. Bajaj, R. van Eldik, L.R. Nevinger, J.B. Keister, Organometallics 8 (1989) 2903
[39] J.B. Keister, U. Frey, D. Zbinden, A.E. Merbach, Organometallics 10 (1991) 1497
[40] W.J. le Noble (Ed.): Organic High Pressure Chemistry, Elsevier, Amsterdam (1988)
[41] A. Rahm, Chapter 10 in ref. [40]
[42] J. Jurczak, A. Rahm, in High Pressure Chemical Synthesis, J. Jurczak and B. Baranowski (Eds), Elsevier, Amsterdam, Chapter 11 (1989)
[43] K.H. Dötz, in Transition Metal Carbene Complexes, Verlag Chemie, Weinheim, 191 (1983)
[44] Y. Yamamoto, K. Maruyama, K. Matsumoto, J. Am. Chem. Soc. 105 (1983) 6963
[45] J. Jurczak, A. Golebiowski, A. Rahm, Tetrahedron Lett. 27 (1986) 853
[46] A. Rahm, F. Ferkous, J. Jurczak, A. Golebiowski, Synth. React. Inorg. Met. Org. Chem. 17 (1987) 937
[47] W.G. Dauben, J.M. Gerdes, G.C. Look, Synthesis (1986) 532
[48] Y. Yamamoto, K. Maruyama, K. Matsumoto, Organometallics 3 (1984) 1583

Cleavage of the Carbon-Hydrogen Bond on Achiral and Chiral Transition Metal Complexes

Lutz Dahlenburg

Institut für Anorganische Chemie
der Friedrich-Alexander-Universität Erlangen-Nürnberg
Egerlandstraße 1, D-8520 Erlangen

Summary

Inter- and intramolecular activation of carbon–hydrogen bonds has been investigated by reduction of ruthenium complexes L_4RuCl_2 containing tri- and tetradentate polyphosphines as ancillary ligands. The transient cis-unsaturated 16e fragments $L_4Ru(0)$ (L_4 = $P(CH_2CH_2CH_2PMe_2)_3$, "pp_3"; $N(CH_2CH_2PMe_2)_3$, "np_3"; $N(CH_2CH_2PPh_2)_3$, "np'_3"; $PMe_3/MeSi(CH_2PMe_2)_3$, "sip_3"), so generated, exhibit a distinct selectivity with respect to insertion across aromatic C–H bonds; yet they also undergo competitive cyclometalation. Reduction of $(np_3)RuCl_2$ with Na(Hg) in refluxing benzene thus affords a 4:1 mixture of $[(Me_2PCH_2CH_2)_2NCH_2CH_2P(Me)CH_2]RuH$ and $(np_3)RuH(C_6H_5)$, but on decreasing the temperature the intermolecular insertion of $(np_3)Ru(0)$ begins to occur preferentially to the cyclometalation reaction. In their reactions with substituted arenes, the 16e equivalents $(pp_3)Ru(0)$ and $(sip_3)(Me_3P)Ru(0)$ show a clear-cut preference for insertion across the unhindered meta- and para-C–H bonds. In no case is activation of the weak benzylic C–H bonds observed. In contrast, reduction of $(pp_2)(Me_3P)RuCl_2$ ("pp_2" = $MeP(CH_2CH_2CH_2PMe_2)_2$) in mesitylene produces, via $(pp_2)(Me_3P)Ru(0)$ and $(pp_2)Ru(0)$ as viable intermediates, the product of benzyl C–H bond cleavage, $(pp_2)(Me_3P)RuH(CH_2C_6H_3Me_2\text{-}3,5)$, accompanied by $(pp_2)(Me_2PCH_2)RuH$. – The complexes $L_4RuH(C_6H_5)$ (L_4 = pp_3, np'_3, and sip_3/PMe_3) all serve as catalysts for the dimerization of alk-1-ynes RC≡CH (R = Ph, n-Bu, t-Bu), which proceeds to yield 1,4-disubstitued 1-en-3-ynes, RCH=CHC≡CR, with varying E/Z stereoselectivities. Reaction of $(sip_3)(Me_3P)RuH(C_6H_5)$ with $4\text{-}MeC_6H_4NCE$ (E = O, S) leads to insertion of the heteroallenes into the Ru–H bond, generating $(sip_3)(Me_3P)Ru[N(Tol)\text{-}C(O)H]C_6H_5$ and $(sip_3)(Me_3P)Ru[SC(=NTol)H]C_6H_5$ as the primary products. While the latter forms $(sip_3)Ru[N(Tol)C(S)H]C_6H_5$ by loss of PMe_3, the former eliminates $Me_3P{=}NC_6H_4Me\text{-}4$ by reaction with another molecule of isocyanate, affording the acyl compound $(sip_3)Ru[N(Tol)C(O)H]C(O)C_6H_5$ as the ultimate product. – The 14e fragments $[(\pm$ or $-)$-trans-1,2-$C_5H_8(PCy_2)_2]Pt(0)$, generated from enantiomerically pure [trans-1,2-$C_5H_8(PCy_2)_2]Pt(H)\text{-}CH_2CMe_3$ by thermolysis, interact with atropisomeric biaryls to produce, with thus far poor optical yields, C–H-activated aryl hydrides of composition [trans-1,2-$C_5H_8(PCy_2)_2]PtH(biar)$; biarH = $2\text{-}Me_3CC_6H_4{-}C_6H_4CMe_3\text{-}2'$ or $2\text{-}MeC_{10}H_6{-}C_{10}H_6Me\text{-}2'$.

Background

The vast majority of established C—H bond activation reactions occurring at transition metal centers may be broadly classified into two groups according to their reaction pathways: oxidative addition and heterolytic activation, homolytic fission of R—H by M·, such as reported in [1], being rare. Heterolytic C—H bond activation (for obvious reasons also termed "σ metathesis" [2]) involves attack of an electrophilic central metal M, usually of configuration d^0, and a basic ligand X, frequently an alkyl residue R', onto R—H to afford, via a four-centered polarized transition state, a new organometallic product M—R together with a new hydrocarbon molecule R'—H (eq. 1). Oxidative addition reactions proceed at the metal centers of coordinatively unsaturated low-valent complexes L_nM, transforming R—H molecules, on a trajectory as depicted in eq. 2, into a set of two different ligands R^- and H^-.

$$L_nM\text{-}R' + R\text{-}H \longrightarrow \begin{matrix} R'\cdots H \\ \vdots \quad\ \ \vdots \\ L_nM\cdots R \end{matrix} \longrightarrow L_nM\text{-}R + R'\text{-}H \qquad (1)$$

$$L_nM + H\text{-}R \longrightarrow L_nM\overset{H}{\cdots\cdots}R \longrightarrow L_nM{<}^{H}_{R} \qquad (2)$$

It is observed that M—C and M—H bond formation via oxidative addition is induced only by metal species that meet particular structural prerequisites. The minimum requirements for making of the two M—H and M—R bonds are two frontier orbitals on the L_nM fragment: one low-lying vacant of σ type and one filled of π symmetry. This situation, considered to be highly favorable for C—H bond oxidative addition, applies to constrained $d^8 ML_4$ fragments which owing to the stereochemical characteristics of their auxiliary ligands L cannot relax toward their preferred square-planar ("<u>trans</u>-unsaturated") ground-state geometry but are forced to adopt an angular ("<u>cis</u>-unsaturated") high-energy structure. Other fragments also suited for C—H oxidative addition include species with frontier orbitals similar to those of <u>cis</u>-$d^8 ML_4$, such as square-pyramidal $d^6 ML_5$, bent $d^{10} ML_2$, and T-shaped $d^8 ML_3$; Fig. 1. All these are isolobal with methylene, and indeed carbene-like reactivity with a wide range of C—H bonds in organic substrates has been observed for many of them [3].

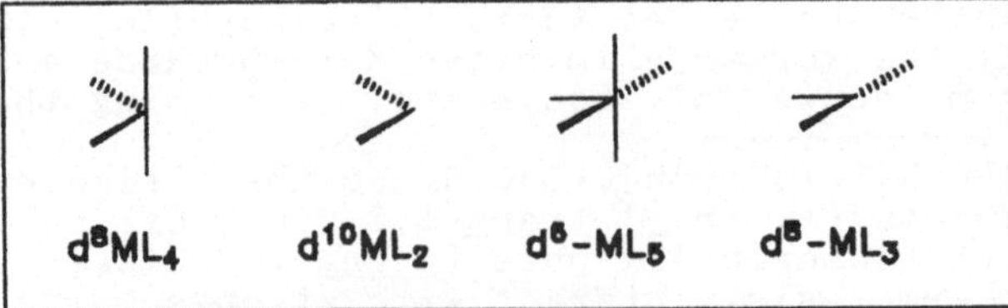

Fig. 1. Isolobal carbene-like complex fragments

Since properly designed polydentate phosphine ligands may be utilized advantageously to change the usual coordination geometry of metal complexes by judiciously selecting parameters such as "relative stereochemical positions of the donor atoms" or "connectivity and number of atoms in the bridging backbones", we synthesized several key complexes having a central ruthenium atom coordinated by tripodal-tridentate and -tetradentate phosphines as given below with which to enter into the C–H oxidative addition chemistry of cis-$L_4Ru(0)$ fragments.

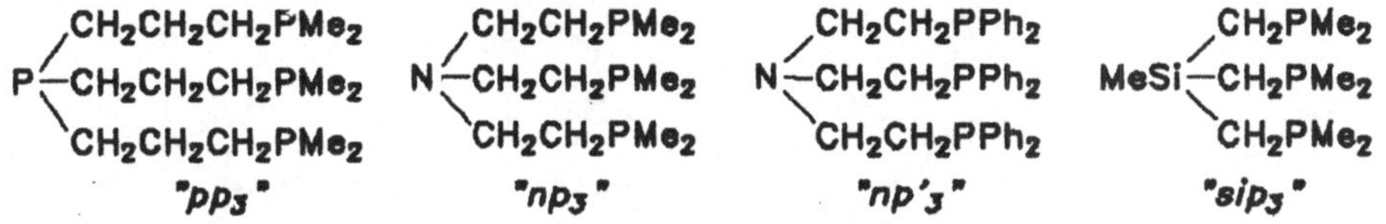

$P(CH_2CH_2CH_2PMe_2)_3$ "pp3"

$N(CH_2CH_2PMe_2)_3$ "np3"

$N(CH_2CH_2PPh_2)_3$ "np'3"

$MeSi(CH_2PMe_2)_3$ "sip3"

Inter- and Intramolecular Insertion Reactions of (np3)Ru(0)

As the cyclometalation of ligand C–H bonds has frustrated many attempts to cleave C–H bonds of non-coordinated hydrocarbons, the importance of fully understanding the factors that govern the competition between intramolecular ligand metalation and intermolecular C–H activation has been generally appreciated. It has been argued that the balance between intra- and intermolecular C–H metalation is controlled by two opposing kinetic effects: (1) conformational strain within the ligand sphere, retarding the cleavage of ligand C–H bonds due to an extra amount of activation energy required to bend the ligand over in the proper orientation for C–H additon to occur; (2) steric congestion about the central metal, which prevents additional molecules from entering into the coordination sphere but may bring coordinated ligands a bit closer to the central metal and could therefore be responsible for the prospensity of many crowded systems to cyclometalate [4]. Frequently however, the intermolecular addition reaction is also made unfavorable for thermodynamic reasons because, for the low-valent late transition metals, M–C bonds are usually not strong enough to offset the loss of entropy associated with reaction 2 [5].

Both from a kinetic and a thermodynamic point of view, the geometrically constrained 16e fragment (np3)Ru(0) appears to make a suitable system for insertion reactions across the C–H bonds of free hydrocarbons, as the conformationally strained but sterically unencumbered chelate rings should suppress the cyclometalation but facilitate the approach of solvent molecules. In addition, the low trans-bond weakening influence of N-bonded ligands should provide the necessary enthalpic driving force for the intermolecular reaction by permitting the formation of a sufficiently strong M–C bond trans to the bridging nitrogen atom. Interestingly however, the fragment (np3)Ru(0), in situ generated by reduction of (np3)$RuCl_2$ with Na(Hg) in C_6H_6 at reflux temperature, enters into a reaction manifold resulting in the production of a 1:4 mixture of the intermolecularly formed phenyl hydride (np3)RuH(C_6H_5) and the intramolecularly closed metalacycle $[(Me_2PCH_2CH_2)_2NCH_2CH_2P(Me)CH_2]RuH$ [6]. Importantly, the balance between intramolecular cyclization and intermolecular hydrocarbon activation can be tipped in favor of the intermolecular reaction by decreasing the temperature: slightly above 25 °, the metalacyclic complex and the product resulting from benzene C–H cleavage are formed in an approximate ratio of 1:1, whereas at 10 to 15 °C, the only compound obtained is $(np_3)RuH(C_6H_5)$:

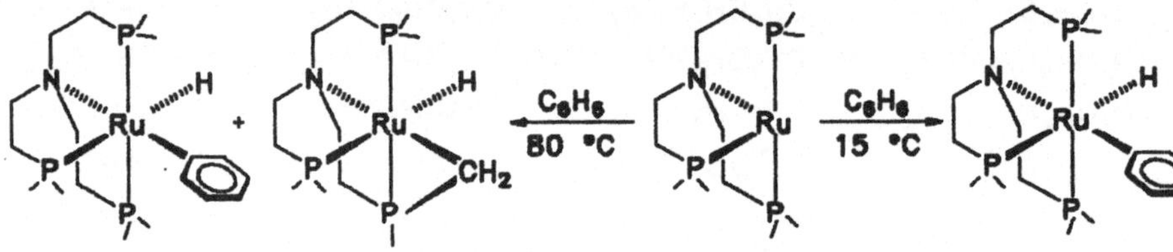

(3)

That lowering the temperature can favor intermolecular C–H activation over cyclometalation has independently been demonstrated for the closely related Rh(I) cation $[(np'_3)Rh]^+$ [7] and also for the pentacoordinate rhenium(I) species $Cp(Me_3P)_2Re$ [8]. In contrast to the rhenium system, where operating at low temperature provides the alkyl hydrides $Cp(Me_3P)_2ReH(R)$ resulting from intermolecular C–H addition as products of kinetic rather than thermodynamic control [8], the preferential formation of $(np_3)RuH(C_6H_5)$ at decreased temperatures reflects the contribution of the $-T\Delta S$ term to the free enthalpy of reaction 3 in that heating the phenyl hydride in solution does not cause reductive elimination of C_6H_6 with concomitant cyclometalation of a PCH_2–H moiety. Rather, the slow formation of $(np_3)RuH(C_6H_5)$ is observed by NMR upon equilibration of the metalacyclic derivative in benzene at ambient conditions.

Insertion of $(pp_3)Ru$ and $(sip_3)(Me_3P)Ru(0)$ across Aromatic C–H Bonds

The 16e fragment $(pp_3)Ru(0)$ which may be generated by photolytic extrusion of dihydrogen from $(pp_3)RuH_2$ [9] as well as by sodium amalgam reduction of $(pp_3)RuCl_2$ [10] exhibits an even greater ease than $(np_3)Ru(0)$ to oxidatively add C–H bonds of arene hydrocarbons. The molecular structures of most of the arylruthenium derivatives $(pp_3)RuH(Ar)$ so accessible (Ar = C_6H_5 [9], C_6H_4Me-3 [10], C_6H_4OMe [11], $C_6H_4NMe_2$ [11], $C_6H_4CF_3$-3 [12], $C_6H_3Me_2$-3,4 [10], $C_6H_3Me_2$-3,5 [10]) have been determined by X-ray diffraction. Because of large steric interactions of the two opposite Me_2P substituents of the pp_3 ligand with the aromatic rings, rotation of the latter about the Ru–C bond is associated with very high barriers, which readily accounts for the isolation of the tolyl hydride as a mixture of two non-interconvertible isomers, containing the <u>meta</u>-methyl substituent oriented away from, or in direction of, the Ru–H linkage (Fig. 2).

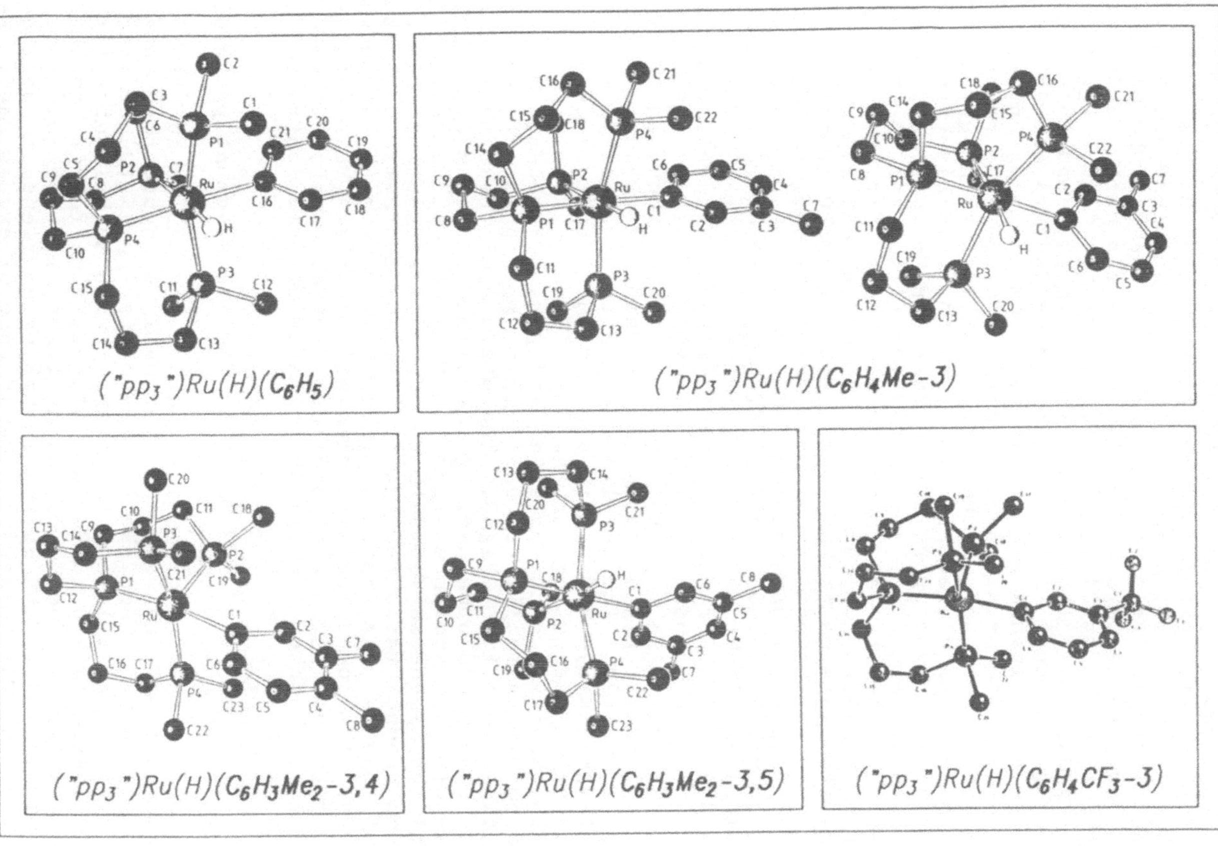

Fig. 2. Molecular structures of several aryl hydrides ("pp3")RuH(Ar)

Initially, it was assumed that the outcome of these metalation reactions reflected a distinct preference of the $(pp_3)Ru(0)$ fragment for insertion across meta-C–H bonds of substituted aromatics [10]. This was eventually shown to be false, at least for $(sip_3)(Me_3P)Ru(0)$, a closely related analogue of $(pp_3)Ru(0)$. Metalation of benzene by this intermediate affords the phenyl hydride $(sip_3)(Me_3P)RuH(C_6H_5)$, the ^{13}C NMR spectrum of which reveals separate signals for the six carbon atoms in the C_6H_5 ring, δ = 174 (ipso), 153.2 (ortho), 146.3 (ortho'), 125.0 (meta), 124.6 (meta'), and 119.9 (para), and thus gives convincing evidence that in $(sip_3)(Me_3P)RuH(Ar)$ derivatives, too, the aryl ligands are prevented from rotating unrestrictedly about the metal–carbon linkage. Since the resonances originating from the two chemically non-equivalent ortho-carbon atoms are all characterized by exceptional high-frequency shifts as well as by partially resolved P,C-coupling in their high-field signals, they are easily identified and, hence, may be used conveniently as spectroscopic tools for probing the number of isomers formed in oxidative C–H addition reactions of substitued arenes to the 16c $(sip_3)(Me_3P)Ru(0)$ fragment. As can be seen by inspection of the results obtained for a series of xylyl and tolyl derivatives (Fig. 3), the single product obtained from oxidative addition of meta-xylene is the expected $(sip_3)(Me_3P)RuH(C_6H_3Me_2\text{-}3,5)$, formed by attack of the 16e intermediate on the least hindered C–H bond. Metalation of ortho-xylene correspondingly proceeds by cleavage of the unhindered C–H bonds para to either methyl group, producing the two conformationally rigid rotamers of $(sip_3)(Me_3P)RuH(C_6H_3Me_2\text{-}3,4)$ in 1:1 ratio. The tolyl and α,α,α-trifluorotolyl derivatives $(sip_3)(Me_3P)RuH(C_6H_4CH_3)$ and $sip_3(Me_3P)RuH(C_6H_4CF_3)$ are particularly interesting in that the three different isomers in which the least hindered meta- and para-C–H bonds have been activated are actually produced in a meta:para:meta product ratio close to 1:1:1; Fig 3. The mixture of these three complexes is obtained upon heating the dichloro precursor (sip_3)-$(Me_3P)RuCl_2$ with sodium amalgam in toluene, α,α,α-trifluorotoluene, or xylene solvents to ca. 80 °C, suggesting that the observed product ratio represents the thermodynamic distribution of isomers. As reported previously, both meta- and para-C–H bonds of toluene are also cleaved by $(C_5Me_5)(Me_3P)Rh(I)$ to afford meta- and para-tolyl hydrides $(C_5Me_5)(Me_3P)RhH(C_6H_4Me)$ in a 2:1 ratio under conditions of thermodynamic control [13].

Activation of Benzylic C–H Bonds by the 14e Fragment $pp_2Ru(0)$

In no case (even with para-xylene or mesitylene as substrates) was activation of the weak and sterically unencumbered benzyl C–H bonds observed when the 16e fragments $(pp_3)Ru(0)$ or $(sip_3)(Me_3P)Ru(0)$ were allowed to interact with methyl-substituted arenes heated at 60 to 90 °C. Although the former could be trapped as $(pp_3)Ru(PMe_3)$ by adding trimethylphosphine to mixtures of $(pp_3)RuCl_2$ and Na(Hg) in hot para-xylene, no well defined product was isolated in the absence of excess PMe_3. Reduction of $(sip_3)(Me_3P)RuCl_2$ in mesitylene or para-xylene did however result in clean conversion to the expected product of

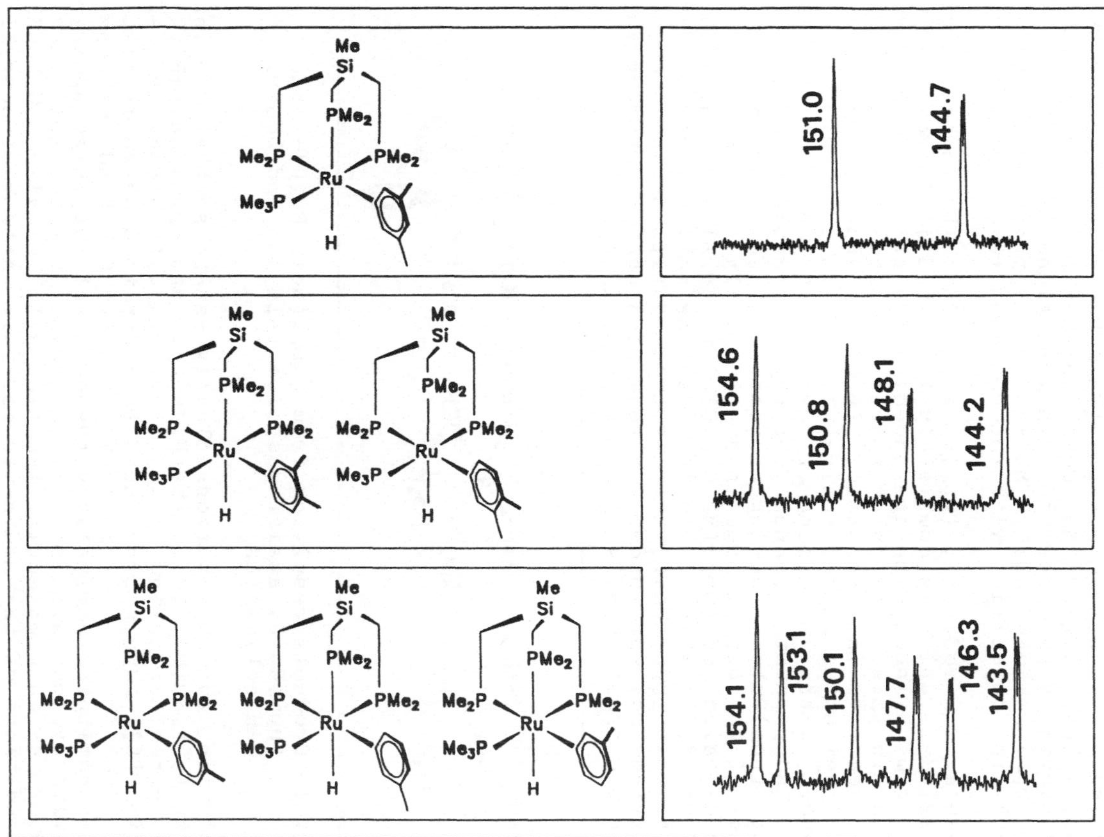

Fig. 3. ^{13}C NMR assignment of meta- and para-metalated structures to isomeric tolyl and xylyl hydrides $(sip_3)(Me_3P)RuH(Ar)$

cyclometalation, $(sip_3)(Me_2PCH_2)RuH$ [6].

Interestingly, reduction of $(pp_2)(Me_3P)RuCl_2$ (containing the open-chain "pp_2" ligand $MeP(CH_2CH_2CH_2PMe_2)_2$ instead of the branched-chain "sip_3" phosphine) in mesitylene at 90 °C gives rise to $(pp_2)(Me_3P)RuH(CH_2C_6H_3Me_2\text{-}3,5)$ as a product of benzylic attack, in competition with the compound resulting from intramolecular C–H addition, $(pp_2)(Me_2PCH_2)RuH$. The formation of the benzyl derivative is completely inhibited if large quantities of trimethylphosphine are added, indicating that product formation in this particular system involves at least two distinct reactive intermediates, $(pp_2)(Me_3P)Ru(0)$ and (pp_2)-Ru(0), of which the latter activates benzylic sp^3-C–H bonds in preference to the methylene or methyl C–H bonds of its own ligand. Independent evidence for the intermediacy of a 14e fragment, $(pp_2)Ru(0)$, having a pronounced tendency to cleave C–H bonds of the benzyl type intermolecularly arises from the clean conversion of $(pp_2)(Me_3P)RuCl_2$ to $(pp_2)(4\text{-}MeC_5H_4N)RuH$-$(CH_2C_6H_3Me_2\text{-}3,5)$ which is isolated as the only product of C–H cleavage when $(pp_2)(Me_3P)RuCl_2$ is treated with sodium amalgam in mesitylene solvent under a stream of argon (to remove any PMe_3) in the presence of 4-picoline; see Fig 4.

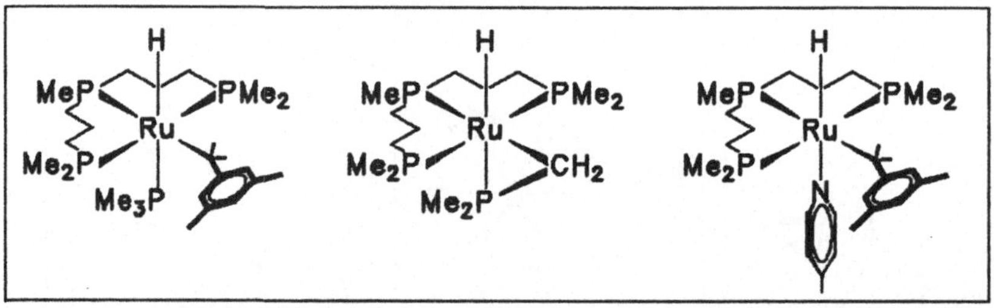

Fig. 4. Products resulting from $(pp_2)(Me_3P)Ru(0)$- and $(pp_2)Ru(0)$-assisted cyclometalation and benzyl C–H activation

In conclusion, our observations clearly establish that the 16e fragments $(pp_3)Ru(0)$, $(sip_3)(Me_3P)Ru(0)$, and $(pp_2)(Me_3P)Ru(0)$ (which, parenthetically, also interacts with arenes by Ar–H activation [6]) have a pronounced tendency to undergo intermolecular insertion across the unhindered <u>sp^2</u> C–H bonds of benzene and substituted derivatives thereof; yet they also undergo competitive cyclometalation which occurs to the complete exclusion of any intermolecular reaction with benzylic C–H bonds under conditions of thermodynamic control. Attack of benzylic C–H bonds by the ruthenium atom does however occur if the metal becomes integrated into a 14e intermediate d^8ML_3, probably possessing a T-shaped groundstate geometry [14]. This structure is tolerated by flexible open-chain tridentate phosphines such as $MeP(CH_2CH_2CH_2PMe_2)_2$ which can bond its donor atoms in a meridional triligate fashion to the central atoms of monomeric complexes, but not by branched-chain tridentates like

$MeSi(CH_2PMe_2)_3$, where formation of two connected six-membered chelated rings with trans-arranged phosphorus atoms would create too much strain.

Reactions of Aryl Hydrides with Potentially Functionalizing Reagents

If one could find a small unsaturated molecule that can react with a transition metal compound derived from C—H activation to form an insertion product, elimation would produce a new organic material. The overall result would be the functionalization of a carbon—hydrogen bond, either stoichiometric in that the metal is left in a different form following reaction, or catalytic in that the transition metal complex is restored in its original composition. In examining potential C—H functionalization schemes, we studied the reactions of selected aryl(hydrido)ruthenium(II) complexes with alkynes, isocyanates, isothiocyanates, carbon dioxide, and carbon disulfide.

Reactions of the complexes $L_4RuH(C_6H_5)$ (L_4 = pp_3, np'_3, sip_3/PMe_3) with alk-1-ynes RC≡CH (R = C_6H_5, n-C_4H_9, t-C_4H_9) did not result in conversion of the starting materials to insertion products. Instead, the metal complexes were all found to serve as catalysts for the dimerization of the alkynes, affording 1,4-disubstituted 1-en-3-ynes, RCH═CH—C≡CR, with varying E/Z stereoselectivities [12]; Table 1.

In discussing possible catalytic cycles, there are two important observations to be taken into account in addition to the catalyst- and substrate-dependent E/Z ratios found for the alkyne homodimers:

(1) Attempted equilibration of ruthenium complexes A - C (see Table 1) in C_6D_6, C_6H_5Me, $C_6H_5CF_3$, C_6H_5CN, and PhC≡CH/C_6H_{12}, respectively, does in no case result in reductive elimination of C_6H_6 and re-addition of solvent molecules to form new C—H insertion products.

(2) Stoichiometric reactions of $(pp_3)RuH(C_6H_5)$ with $C_6H_5C≡CD$ in 1:2 molar ratio proceed according to eq. 4 to give (pp_3)-$RuD(C_6H_5)$ together with a partially deuterated 1-en-3-yne homodimer, $C_6H_5C(D/H)═C(D/H)—C≡CC_6H_5$, characterized by averaged D/H ratios of ca. 0.4/0.6 at C-1 and 0.6/0.4 at C-2, respectively.

$$2\ C_6H_5C{\equiv}CD + (pp_3)Ru(H)(C_6H_5) \longrightarrow C_6H_5C{\equiv}C{-}C(D/H)(=0.6/0.4){=}C(D/H)(=0.4/0.6){-}C_6H_5 + (pp_3)Ru(D)(C_6H_5) \quad (4)$$

Table 1. E/Z Stereoselectivities and Yields of Alkyne Dimerizations Catalyzed by $(pp_3)RuH(C_6H_5)$ (A), $(np_3)RuH(C_6H_5)$ (B), and $(sip_3)(Me_3P)RuH(C_6H_5)$ (C) [a)]

Alkyne/Catalyst	Product	Yield
$C_6H_5C{\equiv}CH$/A	$C_6H_5CH{=}CH{-}C{\equiv}CC_6H_5$ E : Z ≈ 95 : 5	65%
$C_6H_5C{\equiv}CH$/B	$C_6H_5CH{=}CH{-}C{\equiv}CC_6H_5$ E : Z > 95 : 5	71%
$C_6H_5C{\equiv}CH$/C	$C_6H_5CH{=}CH{-}C{\equiv}CC_6H_5$ E : Z ≈ 71 : 29	65%
n-$C_4H_9C{\equiv}CH$/A	n-$C_4H_9CH{=}CH{-}C{\equiv}CC_4H_9$-n E : Z < 5 : 95	42%
n-$C_4H_9C{\equiv}CH$/B	n-$C_4H_9CH{=}CH{-}C{\equiv}CC_4H_9$-n E : Z < 5 : 95	72%
n-$C_4H_9C{\equiv}CH$/C	n-$C_4H_9CH{=}CH{-}C{\equiv}CC_4H_9$-n E : Z ≈ 50 : 50	72%
t-$C_4H_9C{\equiv}CH$/A	t-$C_4H_9CH{=}CH{-}C{\equiv}CC_4H_9$-t E : Z > 95 : 5	33%

[a)] Benzene solutions (50 °C; [Ru] = $4{-}6\cdot10^{-2}$ $mol\cdot l^{-1}$, [RC≡CH] = 0.2–0.4 $mol\cdot l^{-1}$; 12 to 48 h)

The most direct mechanistic hypothesis for the catalytic loop involving alkyne C–H addition processes to coordinatively unsaturated $L_4Ru(0)$ intermediates can be safely ruled out for the $L_4RuH(C_6H_5)$ catalysts in view of their inertness toward reductive elimination and re-addition of C–H bonds. An alternative way to open the necessary coordination site would be by initial dissociation of a donor atom of the ancillary ligands. This would create an $L_3(C_6H_5)RuH$ intermediate which then could undergo addition of the first alkyne molecule giving $L_3(C_6H_5)$-$RuH_2(C{\equiv}CR)$. For further reaction, any mechanism involving insertion of the second alkyne molecule across a Ru–H bond seems highly unlikely in that it would convert the first formed $L_3(C_6H_5)RuH(D)(C{\equiv}CC_6H_5)$ intermediate of the labelling reaction either into $L_3(C_6H_5)Ru(D)(CD{=}CHC_6H_5)(C{\equiv}CC_6H_5)$ or into $L_3(C_6H_5)$ $Ru(H)(CD{=}CDC_6H_5)(C{\equiv}CC_6H_5)$, both of which would then undergo reductive elimination, producing deuterated 1-en-3-yne derivatives, (E)-$C_6H_5CH{=}CDC{\equiv}CC_6H_5$ and (E)-$C_6H_5CD{=}CDC{\equiv}CC_6H_5$, respectively, with distributions of H and D that were not observed.

A more likely third alternative involves 1—>3 hydrogen transfer in $L_3(C_6H_5)RuH_2(C{\equiv}CR)$ to generate a vinylidene species, $L_3(C_6H_5)RuH({=}C{=}CHR)$, disposed for 1,2 addition of a second molecule of RC≡CH across its Ru=C double bond. The 1-en-3-yn-

2-yl complex $L_3(C_6H_5)RuH_2[C(C{\equiv}CR){=}CHR]$ so generated could then eliminate the "head-to-head" alkyne homodimers RCH=CH–C≡CR both as _E_ and _Z_ isomers, as observed, since there is no need for the rearrangement of the proposed seven-coordinate ruthenium(IV) alkinyl hydride into the presumably more stable six-coordinate ruthenium(II) vinylidene intermediate to proceed

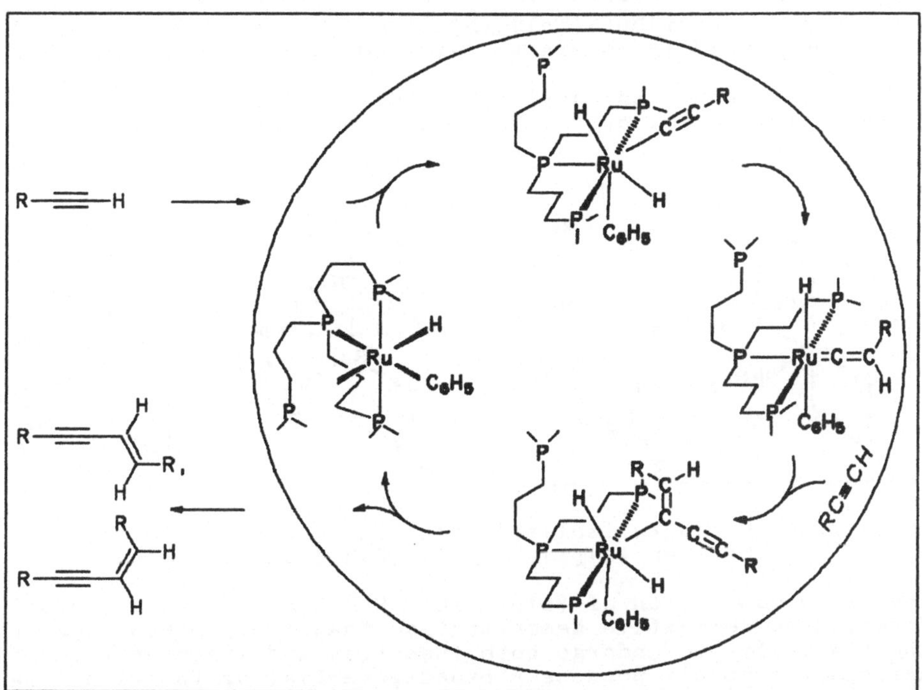

Fig. 5. Catalytic loop proposed for the $L_2RuH(Ar)$-assisted homodimerization of 1-alkynes

stereospecifically. Importantly, such an alkinyl hydride ⟶ vinylidene isomerization would also account for the partial deuteration of the coupling product at both C-1 and C-2, as observed in the stoichiometric labelling experiment, because it would be the hydro as well as the deuterio ligand that could be shifted in $L_3(C_6H_5)RuH(D)(C{\equiv}CC_6H_5)$ to produce a rearranged vinylidene derivative of ruthenium(II). Further support for the involvement of intermediates containing 1-en-3-yn-2-yl ligands $-C(C{\equiv}CR){=}CHR$ according to Fig. 5 arises from literature precedents [15] for the formation of such groups in stoichiometric reactions of $C_6H_6C{\equiv}CH$ at ruthenium centers.

Reactions of $(sip_3)(Me_3P)RuH(C_6H_5)$ with CO_2 and CS_2 result in cleavage of both the Ru–H and Ru–C bond to produce carbonato and trithiocarbonato derivatives of composition $(sip_3)(Me_3P)$-$Ru(E_2CE)$, E = O, S. Stoichiometric reactions of $(sip_3)(Me_3P)$-

RuH(C_6H_5) with <u>para</u>-tolyl isocyanate and isothiocyanate are more interesting in that they result in insertion of the two heteroallenes into the Ru–H bond, giving formamidato-<u>N</u> and thioformimidato-<u>S</u> derivatives, (sip$_3$)(Me_3P)Ru[N(Tol)C(O)H]C_6H_5 and (sip$_3$)(Me_3P)Ru[SC(=NTol)H]C_6H_5, as the primary products. While the latter undergoes PMe_3 dissociation to form the chelate complex (sip$_3$)Ru[N(Tol)C(S)H]C_6H_5 as the final compound, the former readily interacts with another molecule of 4-Me-C_6H_4NCO to produce the acyl (sip$_3$)Ru[N(Tol)C(O)H]C(O)C_6H_5 together with the iminophosphorane Me_3P=NC_6H_4Me-4; Fig. 6 [16].

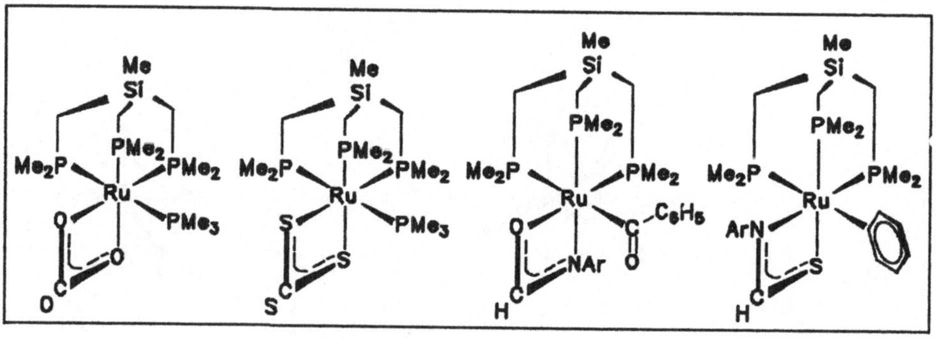

Fig. 6. Products isolated from stoichiometric reactions between (sip$_3$)(Me_3P)RuH(C_6H_5) and heteroallenes

The clean conversion of (sip$_3$)(Me_3P)RuH(C_6H_5) into a benzoylformamidato derivative demonstrates that isocyanates, due to their ability to undergo both insertion and fragmentation reactions, probably possess a broad potential as functionalization groups for the transition metal–hydrogen as well as the transition metal–C bond.

Activation of C–H Bonds under Chiral Conditions

Advances in organometallic chemistry since the early 1980s have shown that dozens of transition metal complexes exhibit a range of reactivity leading to σ bond activation of various hydrocarbons. Virtually no studies have however been reported on the transformation of chiral hydrocarbons by enantiomerically pure metal derivatives, which is rather surprising given the tremendous potential for the application of metal complexes for the functionalization of carbon–hydrogen bonds and the generally appreciated significance of enantioselective organometallic synthesis.

In establishing the objective of "asymmetric C–H activation", which clearly is of interest in the entire area of hydrocarbon chemistry, one could in principle take any R–H molecule containing three different residues connected to a tertiary carbon atom and react this molecule under chiral conditions, e.g. in the presence of resolved chiral ligands L*, with any metal complex known for its ability to insert across carbon–hydrogen bonds:

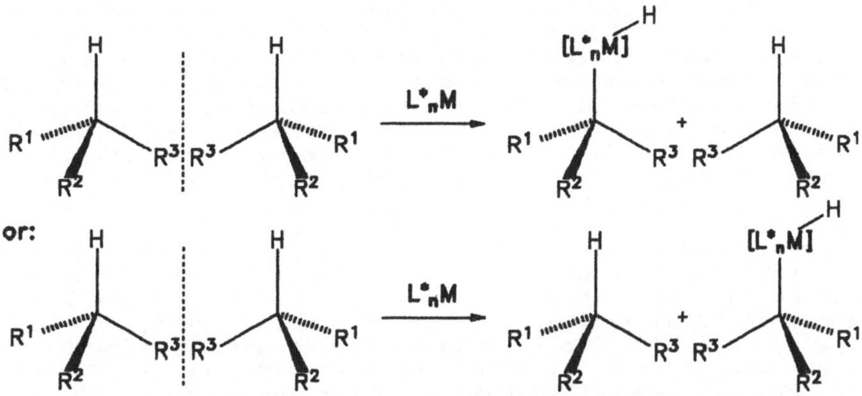

(5)

The selectivity pattern exhibited for different C–H bonds by appropriate metal complexes is however against this approach of kinetically resolving racemic hydrocarbons, since most metal reagents that activate C–H bonds abstract hydrogen from aromatic C–H bonds more readily than from aliphatic ones, and for the latter their ease to become metalated increases in the order 3° < 2° < 1°. The chance of selectively cleaving the tertiary C–H bond at the chiral center of a hydrocarbon molecule such as $C_6H_5C^*H(CH_3)(C_2H_5)$ should hence be thought of as practically nil. Yet, there remain a number of chiral hydrocarbon molecules that are more suited for the purpose of studying C–H activation processes under chiral conditions as the handedness of their optical isomers originates from inherent molecular dissymmetry rather than from the presence of chirality centers at tertiary C atoms. These include ortho,ortho'-disubstituted biaryls which because of their bulky substituents possess a fairly high degree to rotation about their C–C single bonds, e.g.:

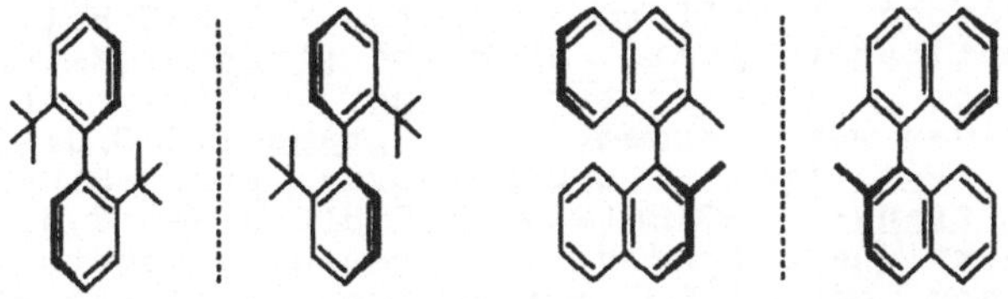

Further support for the use of dissymmetric biaryls as chiral substrates arises from literature precedents giving evidence for the occurrence in arene activation of intermediates that contain the metal coordinated to the arene π system [13]. Obviously, a preliminary step involving substrate association prior to bond cleavage could contribute to enhanced enantiodiscrimination during oxidative C—H addition.

Since for the putative non-linear $d^{10}ML_2$ equivalent (Cy_2PCH_2-CH_2PCy_2)Pt(0), which can be generated via thermolysis of a readily accessible alkylhydrido precursor, ($Cy_2PCH_2CH_2PCy_2$)Pt-H(CH_2CMe_3), a wide range of carbene-like C—H addition reactivity has already been documented [17], it seemed appropriate for initial studies to probe the discriminatory properties of dicoordinate angular platinum(0) species bearing enantiochiral bis(phosphine) ligands. A suitable dissymmetric ditertiary phosphine, closely related to the one already used successfully in $d^{10}ML_2$/R—H activation chemistry (see before), is trans-1,2-cyclopentanediyl[bis(dicyclohexylphosphine)], trans-1,2-$C_5H_8(PCy_2)_2$, which was easily prepared from trans-1,2-C_5H_8-$(PCl_2)_2$ [18] by treatment with cyclohexyl Grignard reagents. Resolution of the racemic mixture so obtained was achieved using optically pure dibenzoyl tartaric acid, DBTA, as a resolving agent [19]. For this purpose, trans-1,2-$C_5H_8(PCy_2)_2$, was first transformed, by oxidation with H_2O_2, to its P,P'-dioxide, trans-1,2-$C_5H_8[P(O)Cy_2]_2$, which was subsequently crystallized as its hydrogen-bridged adduct with DBTA. After cleavage of the PO···DBTA diastereoisomers by aqueous KOH, the degree of resolution, so achieved for the PO enantiomers, was controlled by ^{31}P NMR making use of added chiral lanthanide shift reagents. The enantiomerically pure forms of trans-1,2-$C_5H_8[P(O)Cy_2]_2$, which were isolated after several fractional crystallizations of the diastereoisomeric PO···DBTA adducts, were eventually converted back to their parent trans-1,2-C_5H_8-$(PCy_2)_2$ enantiomers using Ph_2SiH_2 as a reductant [20].

Given the extensive C—H activation chemistry displayed by ($Cy_2PCH_2CH_2PCy_2$)Pt(0) (vide supra), it was not surprising to see that the 14e species [trans-1,2-$C_5H_8(PCy_2)_2$]Pt(0), in situ generated from [(±)-trans-1,2-$C_5H_8(PCy_2)_2$]PtH(CH_2CMe_3) under mild thermal conditions, also oxidatively adds a wide range of sp^3, sp^2, and sp C—H bonds. Thus, thermolysis of the racemic neopentyl hydrido complex in cyclohexane solutions of $C_6H_5C{\equiv}CH$ thus produced an alkynylhydrido complex, [(±)-trans-1,2-C_5H_8-$(PCy_2)_2$]PtH($C{\equiv}CC_6H_5$), in competition with a coordination compound, [(±)-trans-1,2-$C_5H_8(PCy_2)_2$]Pt($HC{\equiv}CC_6H_5$). Thermolysis in benzene proceeded to afford [(±)-trans-1,2-$C_5H_8(PCy_2)_2$]PtH-(C_6H_5) in quantitative yield. When the [trans-1,2-$C_5H_8(PCy_2)_2$] Pt(0) equivalent was generated in meta-xylene, three major platinum hydrides were formed, [(±)-trans-1,2-$C_5H_8(PCy_2)_2$]PtH ($C_6H_3Me_2$-3,5), [(±)-trans-1,2-$C_5H_8(PCy_2)_2$]PtH($C_6H_3Me_2$-2,4), as well as [(±)-trans-1,2-$C_5H_8(PCy_2)_2$]PtH($C_6H_3Me_2$-2,6). Thermolysis in para-xylene afforded mixtures of products resulting from arene and benzyl C—H activation, [(±)-trans-1,2-C_5H_8-$(PCy_2)_2$]PtH($C_6H_3Me_2$-2,5) and [(±)-trans-1,2-$C_5H_8(PCy_2)_2$]PtH-($CH_2C_6H_4Me$-4), respectively. Finally, heating of [(±)-trans-1,2-$C_5H_8(PCy_2)_2$]PtH(CH_2CMe_3) in the presence of hexamethyldisiloxane led to oxidative addition of methyl C—H bonds giving [(±)-trans-1,2-$C_5H_8(PCy_2)_2$]PtH($CH_2SiMe_2OSiMe_3$) [20]:

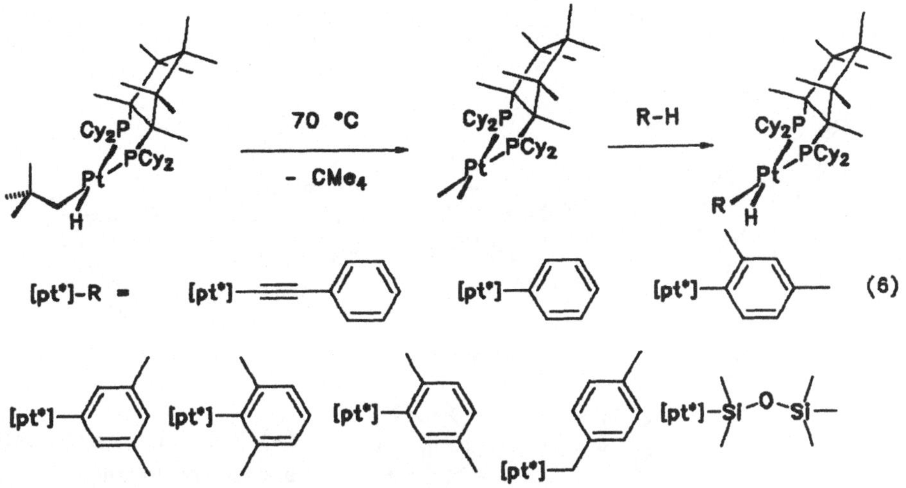

(6)

Thermolyses of the enantiomerically pure starting materials, [(+)-*trans*-1,2-$C_5H_8(PCy_2)_2$]PtH(CH_2CMe_3) or [(−)-*trans*-1,2-C_5H_8 $(PCy_2)_2$]PtH(CH_2CMe_3), were performed at ≈70 °C in cyclohexane solutions containing precisely 2 equivalents of (±)-2-$Me_3CC_6H_4$ $-C_6H_4CMe_3$-2' and (±)-2-$MeC_{10}H_6$—$C_{10}H_6Me$-2', respectively. Monitoring the decomposition of the neopentyl hydrides and the formation of products by ^{31}P NMR spectroscopy indicated the biaryl hydrides [(+)-*trans*-1,2-$C_5H_8(PCy_2)_2$]PtH(biar) and [(−)-*trans*-1,2-$C_5H_8(PCy_2)_2$]PtH(biar) to be produced as isomeric mixtures in quantitative yields. Upon completion of the reactions, the C_6H_{12} solvent was removed by evaporation in vacuo and the unreacted biaryls were isolated from the residues via column chromatography. The degree of optical induction achieved during biar—H oxidative addition was subsequently estimated from the optical rotation of the 2,2'-dimethylbinaphthyl, so recovered, or, in the case of the more volatile 2,2'-di-*t*-butylbiphenyl, from the results of enantioselective gas chromatography on a heptakis(6-*O*-methyl-2,3-di-*O*-pentyl)-ß-cyclodextrin stationary phase [21].

For both substrates, the enantiomeric yields thus far obtained did not exceed 6%. Though certainly not impressive, these optical yields, representing the first enantiomeric excess values to be reported for oxidative C—H addition reactions, are clearly above the level of confidence as it could be shown that the enantiodiscrimination in the resolution of the two biaryls is controlled by the chirality of the ancillary phosphine ligand. Thus, with [(+)-*trans*-1,2-$C_5H_8(PCy_2)_2$]PtH(CH_2-CMe_3) as a source for the chiral intermediate [(+)-*trans*-1,2-$C_5H_8(PCy_2)_2$]Pt(0), the enantiomer (+)-2-$MeC_{10}H_6$—$C_{10}H_6Me$-2' is preferentially enriched in the 2,2'-dimethylbinaphthyl sample recovered from the C—H cleavage reaction; thermolysis of [(−)-*trans*-1,2-$C_5H_8(PCy_2)_2$]PtH(CH_2CMe_3) correspondingly results in a slight enrichment of (-)-2-$MeC_{10}H_6$—$C_{10}H_6Me$-2', the enantiomeric yield remaining the same as with [(+)-trans-1,2-C_5H_8-$(PCy_2)_2$]PtH(CH_2CMe_3) as a resolving reagent [20]:

(+)-Cy_2P PCy_2 Pt H

30 ml C_5H_{12}

68 °C, 1 h

- CMe_4

(+)-Cy_2P PCy_2 Pt (binaph) H

8 mmol 4 mmol (7)

+ unreacted

(+)-2,2'-Dimethylbinaphthyl slightly enriched:
5% < e.e. <6%
(from optical rotation)

Given the ability of alkyl and aryl hydrides of platinum(II) to undergo reductive elimination of R—H at elevated temperature [17], the so far low enantiomeric yields encountered in attempts at the resolution of racemic biaryl dervatives by chiral $d^{10}ML_2$ equivalents may be attributed to a facile thermal equilibration process, interconverting the diastereoisomeric C—H cleavage products [(± or ∓)-trans-1,2-$C_5H_8(PCy_2)_2$]-PtH[biar-(± or ∓)] by reductive elimination and oxidative re-addition of the respective biaryl. For improving the optical yields of the resolution procedure delineated before, alternative routes allowing the enantiomeric [trans-1,2-$C_5H_8(PCy_2)_2$]-Pt(0) coordination complexes to be generated at lower temperature are therefore indicated and will be studied in future work.

Acknowledgments

I wish to thank my collaborators mentioned in the references and especially Dr. A. Saare for her creative contributions to the [$C_5H_8(PCy_2)_2$]Pt(0) chemistry. I also thank Prof. Dr. W.A. König for his continued interest and cooperation. Generous support by the Volkswagen-Stiftung and also by the Fonds der Chemischen Industrie is gratefully acknowledged. Furthermore, I am indebted to DEGUSSA (Hanau) for generous gifts of platinum and ruthenium salts.

References

[1] R.H. Crabtree, J. Chem Educ. 65 (1988) 290.
[2] M.E. Thompson, S.M. Baxter, A.R. Bulls, B.J. Burger, M.C. Nolan, B.D. Santarsiero, W.R. Schäfer, J.E. Bercaw, J. Am. Chem. Soc. 109 (1987) 203.
[3] J.-Y. Saillard in J.A. Davies, P.L. Watson, J.F. Liebman, A. Greenberg (Eds.): Selective Hydrocarbon Activation, Principles and Progress, VCH Publishers, New York 1990, p. 207.
[4] R.H. Crabtree, Chem. Rev. 85 (1985) 245.
[5] P.O. Stoutland, R.G. Bergman, S.P. Nolan, C.D. Hoff, Polyhedron 7 (1988) 1429.
[6] L. Dahlenburg, S. Kerstan, D. Werner, J. Organomet. Chem. 411 (1991) 457.
[7] C. Bianchini, D. Masi, A. Meli, M. Peruzzini, F. Zanobini, J. Am. Chem. Soc. 110 (1988) 6411.
[8] T.T. Wenzel, R.G. Bergman, J. Am. Chem. Soc. 108 (1986) 4856.
[9] M. Antberg, L. Dahlenburg, K.-M. Frosin, N. Höck, Chem. Ber. 121 (1988) 859.
[10] L. Dahlenburg, K.-M. Frosin, Chem. Ber. 121 (1988) 865.
[11] K.-M. Frosin, Dissertation, Universität Hamburg, 1989.
[12] L. Dahlenburg, K.-M. Frosin, S. Kerstan, D. Werner, J. Organomet. Chem. 407 (1991) 115.
[13] W.D. Jones, F.J. Feher, J. Am. Chem. Soc. 106 (1984) 1650.
[14] P. Hofmann, C. Meier, U. Englert, M.U. Schmidt, Chem. Ber. 125 (1992) 353.
[15] A. Dobson, D.S. Moore, S.D. Robinson, M.B. Hursthouse, L. New, Polyhedron 4 (1985) 1119.
[16] C. Becker, Diploma Thesis, Universität Hamburg, 1990.
[17] M. Hackett, J.A. Ibers, G.M. Whitesides, J. Am. Chem. Soc. 110 (1988) 1436; M. Hackett, G.M. Whitesides, J. Am. Chem. Soc. 110 (1988) 1449.
[18] D.L. Allen, V.C. Gibson, M.L.H. Green, J.F. Skinner, J. Bashkin, P.D. Grebenik, J. Chem. Soc., Chem. Commun. 1983, 895.
[19] H. Brunner, W. Pieronczyk, Angew. Chem. 91 (1979) 655; Angew. Chem. Int. Ed. Engl. 18 (1979) 620.
[20] A. Saare, Dissertation, Universität Hamburg, 1991.
[21] W.A. König, D. Icheln, T. Runge, I. Pforr, A. Krebs, J. High Resolution Chromatogr. 13 (1990) 702.

Synthetic Applications of Mercury Photosensitization

Robert H. Crabtree

Yale Department of Chemistry, New Haven, CT. 06520, USA

Summary

A method for the dehydrodimerization of organic compounds is described which relies on Hg photosensitized reactions either under reflux or in the presence of H_2. Radicals formed in the gas phase recombine and disproportionate, but the disproportionation product is returned to the radical pool by H atom addition. The dehydrodimer is protected from further conversion by condensation. Other reactive gases, such as O_2, CO, and SO_2 lead to functionalized products.

1 Introduction

The functionalization of unactivated C-H bonds in organic compounds is an area of current interest [1]. Organometallic catalysts for alkane dehydrogenation suffer from the problem of degradation because of the fragility of the ligands and participation of the solvent [2]. We have been particularly attracted by the Hg photosensitization method because it avoids both ligands and solvent. In the simplest version of the reaction, we reflux the substrate with a drop of Hg in a quartz vessel under 254 nm irradiation. Hydrogen is evolved and the dehydrodimer of the starting alkane is formed (eq. 1).

$$2R\text{-}H \rightarrow R_2 + H_2 \qquad (1)$$

2 Hg* Method

The reaction happens exclusively in the gas phase and it is this which gives the reaction its high selectivity for the dehydrodimer. The initial oxidation product condenses and is protected from further conversion. The Hg emission line of the required low pressure Hg lamp excites only the vapor and not the dissolved Hg, because it matches only the absorbtion spectrum of the vapor phase Hg. We verified the reaction does not go on the surface of the Hg drop or of the wall of the vessel. The reaction zone is a thin ring just within the inside surface of the reactor, where the light first encounters the Hg vapor. Hg, being such a strong absorber of Hg radiation picks up essentially all the light energy and this protects the bulk of the organic substrate from conventional organic photochemistry. Most reactions are run in the range 0°-150° and at atmospheric pressure. Higher boiling species can be run under reduced pressure or by steam distillation; even n-$C_{18}H_{38}$ could be dimerized successfully in these ways. The lower temperatures are most useful when the radical formed is particularly sensitive to thermal decomposition. Quantum yields are in the range 0.1-0.8 and multigram quantities of product are obtainable in an 8W reactor overnight [3].

Although the reaction has been known for decades [4], there are few, if any, organic synthetic applications. The mechanism, worked out by Steacie and others [5], is shown in eq. 2-5. Hg* is the

$^3P_1$ excited state which is produced with 254nm light.

$$Hg + h\nu \rightarrow Hg^* \quad (2)$$

$$Hg^* + RCH_2CH_3 \rightarrow RCH{\cdot}CH_3 + H{\cdot} \quad (3)$$

$$H{\cdot} + RCH_2CH_3 \rightarrow RCH{\cdot}CH_3 + H_2 \quad (4)$$

$$2RCH{\cdot}CH_3 \rightarrow (RCHCH_3)_2 + RCH{=}CH_2 + RCH_2CH_3 \quad (5)$$

The formation of the alkene disproportionation product, $RCH{=}CH_2$ in eq. 5, might be expected to complicate the products, but this is prevented because H atoms produced in eq. 3 rapidly add to the alkene by eq. 6 to return it to the radical pool.

$$RCH{=}CH_2 + H{\cdot} \rightarrow RCH{\cdot}CH_3 \quad (6)$$

This is shown by the fact that cis-1,4-dimethylcyclohexane loses stereochemistry in forming both the dehydrodimer and the alkane disproportionation product, trans-1,4-dimethylcyclohexane, which is now distinguishable from the starting alkane (eq. 7). By running under D_2, we find deuterium in the appropriate places in the dehydrodimer as is consistent with H(D) atom attack of the alkene disproportionation product.

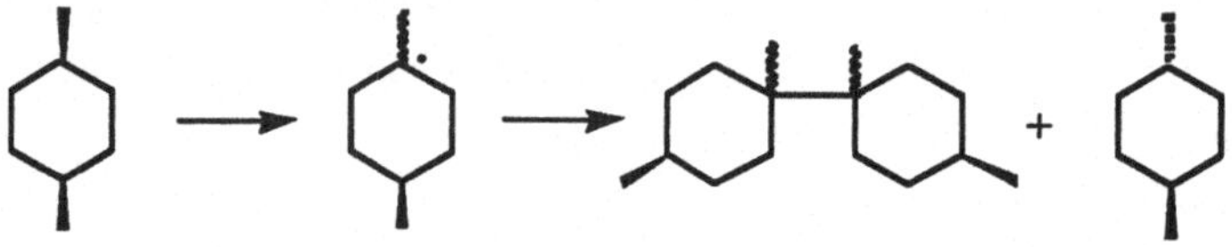

(7)

The regiochemistry is determined by the selective attack of the Hg* on the weakest C-H bonds in the molecule; the radicals then recombine statistically. Where several types of CH bond are present, we can quantitatively account for the product ratios knowing the bondstrengths, or estimate the bondstrengths knowing the product ratios. Where they are possible, meso and dl isomers are formed in equal amounts, although one isomer could sometimes be crystallized cleanly from the mixture.

The reaction is easily extended to alcohols, ethers and silanes, for example, glycols are easy to make from the corresponding alcohols e.g., eq. 8.

$$(CF_3)_2CHOH \rightarrow (CF_3)_2CH(OH)CH(OH)(CF_3)_2 \quad (8)$$

$$Et_2SiH_2 \rightarrow Et_2SiH\text{-}SiHEt_2 \quad (9)$$

Note the formation of the 1,2-dihydrosilane in eq. 9, a result of the condensation of this material before it can react further (vapor pressure selectivity). The homodimer of piperidine can be dehydrogenated over copper chromite to give 2,2'-bipyridyl.

Mixing any two reactive substrates leads to the formation of a mixture of the homo- and hetero-dimers, which in most cases are very easy to separate as a result of their great polarity or volatility differences.

$$RH + CH_3OH \rightarrow R_2 + RCH_2OH + HOCH_2CH_2OH \quad (10)$$

$$Et_3SiH + MeCH_2OH \rightarrow Et_3Si\text{-}CH(Me)OH \quad (11)$$

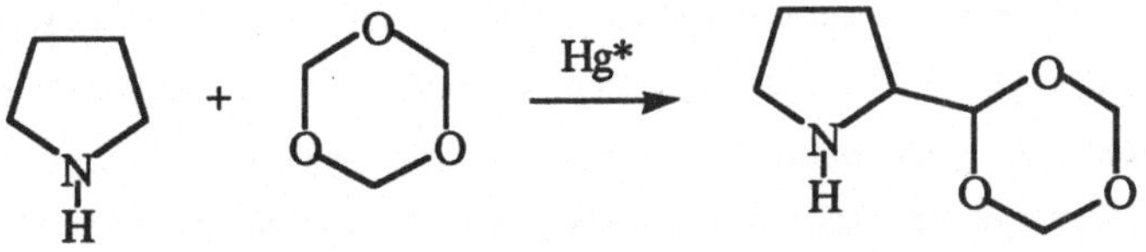

$$\text{pyrrolidine} + \text{trioxane} \xrightarrow{Hg^*} \text{2-(1,3,5-trioxan-2-yl)pyrrolidine} \quad (12)$$

In all other cases, the reaction of a silane and an alcohol forms a product with a Si-O bond, as a result of what is usually the greater kinetic reactivity of the alcohol OH group and the greater thermodynamic stability of a Si-O vs. a Si-C bond. Eq 11 shows how our chemistry leads to the unusual Si-C coupling product. In eq. 12, a protected form of prolinal is formed from formaldehyde (trioxan in the vapor) and pyrrolidine.

3 Hg*/H_2 Method

As shown by Breckenridge and Soep,[6] Hg* binds to the C-H bond of an alkane to form an excited state complex which goes on to give C-H cleavage (eq. 13).

$$Hg^* + H\text{-}C \rightarrow \{Hg^*\text{-}H\text{-}C\} \rightarrow Hg + C\cdot + H\cdot \quad (13)$$

Hg* has the electronic structure $d^{10}s^1p^1$ (compare ground state Hg: $d^{10}s^2p^0$). This means that it can bind to almost any lone pair as well or better than it does to C-H bonds. This is because the lone pair and the Hg(s^1) electron lead to a three electron bond, formally of half order. Hg*-S exciplex formation (S = substrate) is relatively strong, for example, the dissociation energy for Hg*-NH_3 is 17 kcal mol^{-1}. When the binding becomes so strong that the C-H bonds in the molecule are no longer homolyzed, the reaction fails. This is the case for ketones, amides, esters, carboxylic acids, most amines, and alkenes. Arenes are also ineffective; binding in this case may be in the η^6-mode as in **1**.

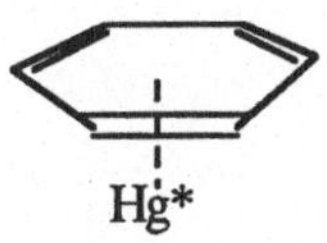

Hg*

1

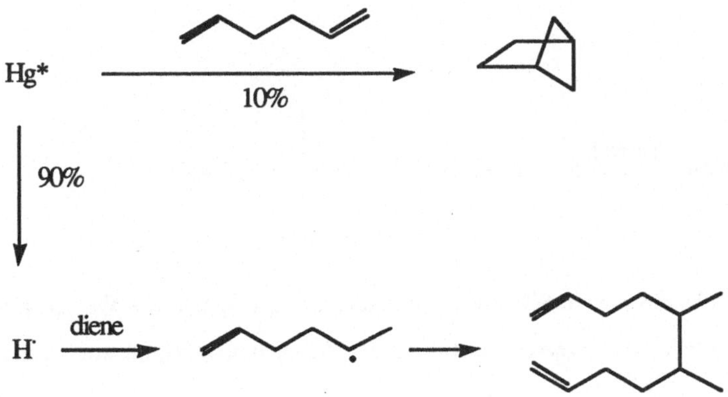

Hg* —(10%)→ [bicyclic product]

Hg* —(90%)→ H· —diene→ [radical] → [dimer] (14)

Molecular hydrogen is the most reactive substrate and gives H atoms, which can abstract H atoms from CH bonds (compare eq. 4). This Hg*/H_2 system is effective in forming the C radicals even in the presence of groups which are incompatible with the Hg* conditions described above. This is because H atoms do not give exciplexes. Refluxing the liquidwould lead to the expulsion of the reactive H_2 gas from the reflux zone and we lower the temperature so that the vapor phase now contains both H_2 and substrate.

Hg*/H 2 (15)

OH Hg*/H 2 OH OH (16)

N O Hg*/H 2 O N N O (17)

NH2 Hg*/H 2 NH2 NH2 (18)

COOH Hg*/H 2 HOOC COOH (19)

Typical pressures of H_2 and substrate are 660 mmHg and 100 mmHg, respectively. We can show that under these conditions essentially all the reaction passes via Hg*/H_2 and not Hg*/substrate reactive collisions by studying the reaction of 1,5-hexadiene, which gives two different products by the two paths (eq. 14). Even at a 50:50 mole ratio of diene and H_2 in the vapor led to a 90:10 ratio of Hg*/H_2 and Hg*/diene products.

Eq. 15-19 shows examples of substrates which were not at all, or only inefficiently dehydrodimerized under Hg* conditions but which now can be used successfully.

Since H atoms add to alkenes, these substrates should be reactive under Hg*/H_2 conditions. With saturated substrates, the new C-C bond is formed where the CH bond was wekest in the original sustrate, therefore usually a to the most activating substituent. The great advantage of the H atom reactions with alkenes is that the position of C-C bond coupling can be defined by a suitable choice of alkene. The H atom adds to form the stablest radical, so from 1-hexene we see largely 5,6-dimethyldecane. The minor products from 1-hexene are 5-methylundecane (10.75%) and n-dodecane (0.25%). These numbers imply that the H atom adds to the more substituted and less substituted ends of the 1-hexene with a selectivity of ca. 5:95.

Of greater synthetic interest are the reactions shown in eq. 20-24, where the alkene bears other functionality. The method is tolerant of functionality such as -CN, -CO_2H, ester, ketone, amide, fluoroalkyl and epoxy groups. In the fluoroalkene case, the two H substituents show the terminal addition of the H atom to the C=C double bond. H atom addition occurs preferentially at the C=C double bond rather than the CN triple bond in eq. 21 or the CO double bond in eq. 22-23. We were very surprised that the epoxy-substituent was not affected in eq. 24. Quantum and chemical yields were satisfactory in all these cases. In a setup useful for more volatile substrates, the hydrogen was presaturated with substrate vapor before being introduced into the photoreactor. Otherwise the substrate was placed in the quartz vessel itself. Disproportionation products tend to be swept out of the reactor in the exit gases. If they are allowed to build up in the reactor, selectivity can suffer.

$$n\text{-}C_4F_9\text{-}CF{=}CF_2 \xrightarrow{Hg^*/H_2} (n\text{-}C_4F_9)(CF_2H)CF\text{-}CF(n\text{-}C_4F_9)(CF_2H) \quad (20)$$

$$CH_2{=}C(CH_3)\text{-}CN \xrightarrow{Hg^*/H_2} (CH_3)_2C(CN)\text{-}C(CN)(CH_3)_2 \quad (21)$$

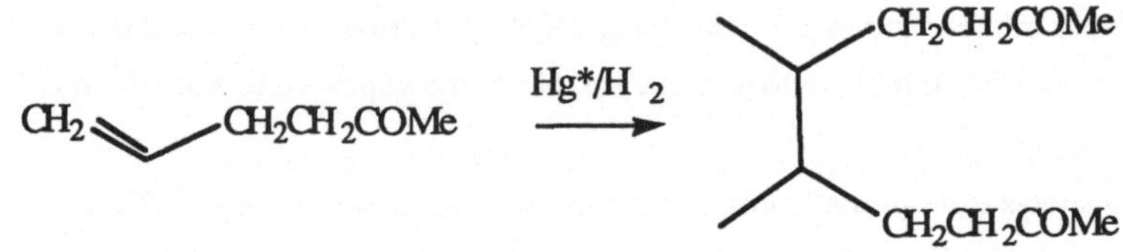
CH2
CH2CH2COMe
Hg*/H 2
CH2CH2COMe
CH2CH2COMe

(22)

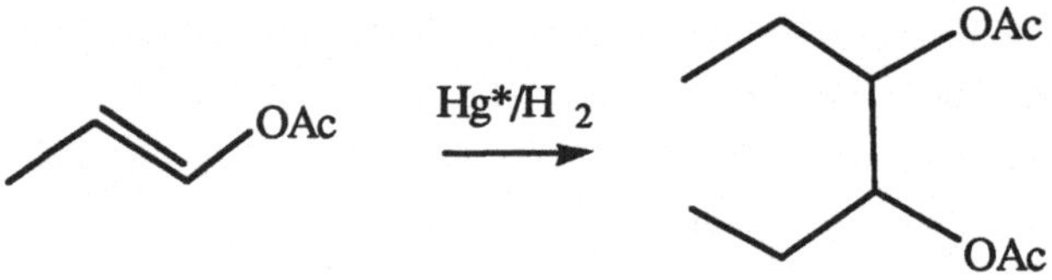
OAc
Hg*/H 2
OAc
OAc

(23)

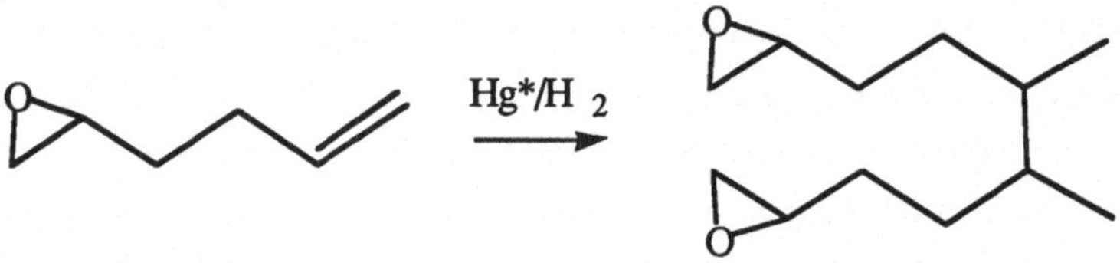
Hg*/H 2

(24)

Radicals can behave differently according to whether they are formed by H atom addition to alkene or H atom abstraction from alkane. We see 5% rearrangement by an unusual 1,2-shift of the methyl group (eq. 25) when tBuCH·CH_3 is formed by H atom addition to tBuCH=CH_2, but no rearrangement when the same radical is formed from the alkane.

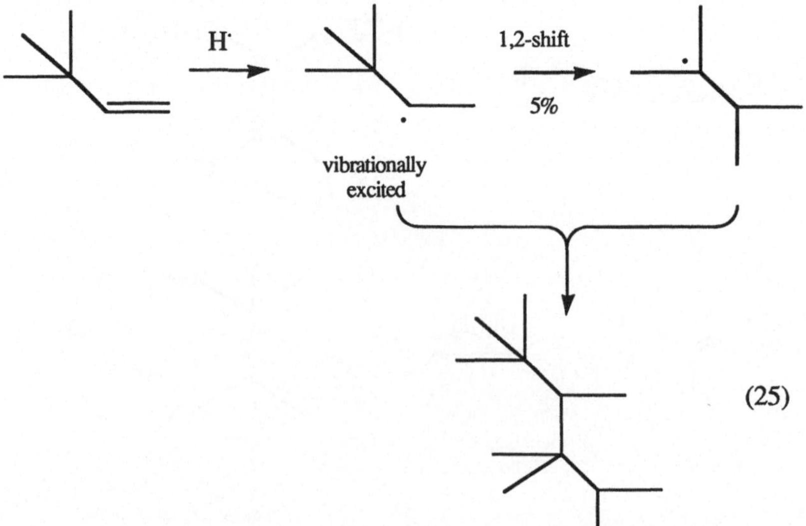
H·
1,2-shift
5%
vibrationally
excited

(25)

Adding CO_2, known to quench vibrational energy efficiently, reduces the amount of rearrangement to 2%, suggesting this is a reaction of the vibrationally 'hot' radical. We estimate that H atom addition to an alkene leaves the new radical with ca. 40 kcal mol^{-1} of thermal energy. In solution this would be quickly quenched by the solvent, but in the vapor the energy lasts long enough for significant rearrangement to take place. At much lower pressures, such as were employed in the early physico-chemical studies, [5] there is extensive radical fragmentation because the excess energy is quenched much less efficiently than here, where we operate at 1 atm. This fragmentation may have discouraged practical applications of the chemistry. Not only is it much easier to operate at 1 atm., but we also largely suppress reactions of hot radicals in this way. In addition, the more complex substrates we use have more degrees of freedom into which the excess vibrational energy may be channeled; this factor also decreases the probablity of fragmentation before the 'hot' radical is collisionally deexcited.

We looked for some of the characteristic cyclization reactions of radicals. H atom addition to a suitable diene, like **2** gives essentially only the cyclized products (eq. 26). The same products are seen from the 1-alkyne of the same carbon number. The proposed mechanism is illustrated in eq. 27, in which the formation of the vinyl radical **3** occurs in the first step. These radicals are much more reactive for H atom abstraction than are alkyl radicals, as illustrated by the bond strength differences between a vinyl C-H (108 kcal mol^{-1}) and a 2° alkyl C-H (95 kcal mol^{-1}). An intramolecular abstraction is likely to be strongly favored entropically. Abstraction leads to the same radical formed in eq. 25 and therefore to the same final products. Aromatic systems also undergo H atom addition, but the product mixtures are complicated and the reaction is not useful.

H·

X X X

2

(26)

X + X X X

X = CH_2, O, NH, $SiMe_2$

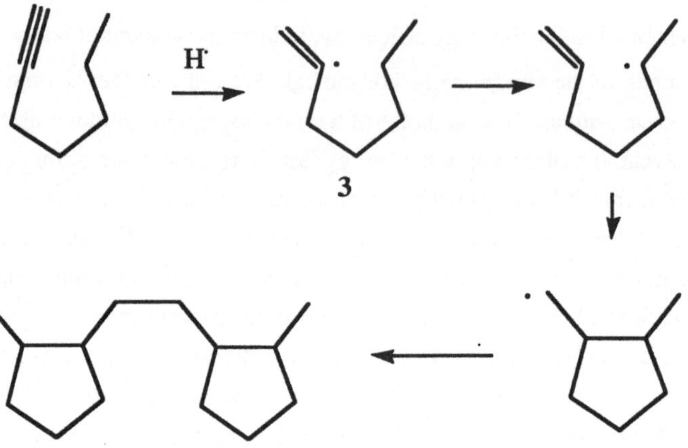

H·

3

(27)

Usually, cross dimerizations involving alkanes (RH) and another substrate (R'H) give R_2, RR' and R'_2 in a statistical recombination. Where R'H is an unsaturated alcohol, such as $Me_2C{=}CHCH_2OH$, essentially only the homodimers R_2 and R'_2 were formed. We propose that this is due to H bonding of the alcohols in the vapor phase, leading to a rate increase for alcohol-alcohol dimerization, as shown in eq. 28. We can show that the rate of alcohol-alcohol recombination is elevated 7000 fold for $Me_2C{=}CHCH_2OH$ and cyclohexene.

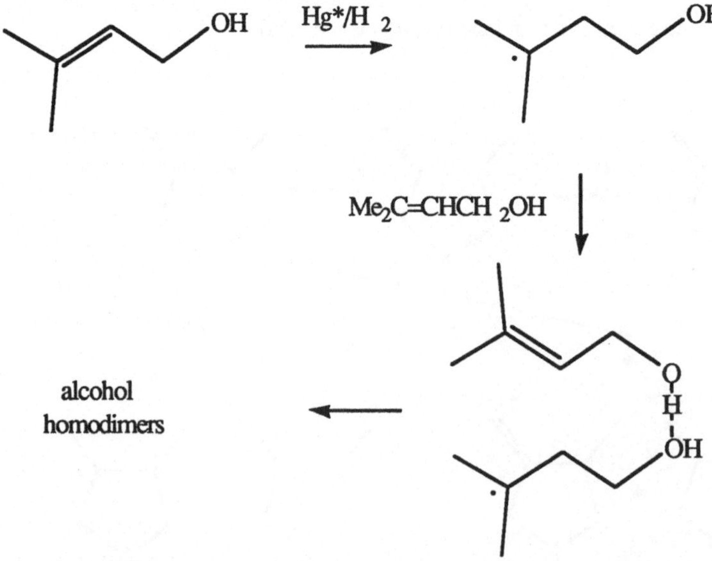

OH Hg*/H_2 OH

$Me_2C{=}CHCH_2OH$ (28)

alcohol homodimers

O H OH

4 Other Reactive Gases

Other reactive gases are effective. For example, with alkanes under Hg*/SO_2 conditions, [7] the sulfur dioxide efficiently traps the radicals formed by Hg*/alkane reactive collisions. This can almost entirely prevent dimerization of the R· radicals and the species shown in eq. 29 are formed instead. Most of the components of this mixture can be oxidized to the sulfonic acid with performic acid. CO is a less efficient trap. Ketones are formed in this case, but dehydrodimer formation was never supressed entirely at the pressures we studied (up to 1 atm).

$$\mathrm{RH} \xrightarrow{\mathrm{Hg^*,\,SO_2}} \underset{14\%}{\mathrm{R_2}} + \underset{18\%}{\mathrm{R_2SO_2}} +$$

$$\underset{16\%}{\mathrm{RSOOR}} + \underbrace{\mathrm{RSOOH} + \mathrm{RSSO_2R} + \mathrm{RSO_3H}}_{\text{61\% (RSOOH and its disproportionation products)}} \quad (29)$$

$$\xrightarrow{\mathrm{HCO_3H}} \mathrm{RSO_3H} \qquad (\mathrm{R = C_6H_{11}})$$

Hg/air gives hydroperoxides even with methane, which normally is resistant to attack by Hg* because of the high C-H bondstrength. Either the Hg* first attacks O_2 to give some excited state or this is a chain reaction. Hg*/N_2O seems to give O atoms and these seem to abstract H, rather than insert into C-H bonds. For example, CH_4 and CD_3OD gave propylene glycol in which the methyl group derives from methane. Initial recombination of $CH_3\cdot$ and $\cdot CD_2OD$ give ethanol which is still volatile under the conditions and so cross dimerizes with the excess methanol to give propylene glycol (eq. 30). [6]

$$\mathrm{CH_4} \xrightarrow[\mathrm{=O\ atom}]{\mathrm{Hg^*/N_2O}} \mathrm{CH_3^{\cdot}} + {}^{\cdot}\mathrm{OH} \xrightarrow{\mathrm{CD_3OD}} {}^{\cdot}\mathrm{CD_2OD}$$

$$\mathrm{CH_3CD(OD)CD_2OD} \xleftarrow{\mathrm{CD_3OD}} \underset{\text{volatile}}{\mathrm{CH_3CD_2OD}} \quad (30)$$

5 Conclusion

Mercury photosensitization provides a way to couple simple organic compounds on a large scale to make a variety of homo- and cross-dimers and to hydrodimerize alkenes in simple and widely available apparatus at 1 atm. pressure and at ambient or reflux temperature. Vapor phase selectivity allows us to obtain the initial products uncontaminated by overoxidation products. Most of the reactions are thermodynamically unfavorable under the conditions used and the photon energy is therefore required to drive the reaction. These conditions are not applicable to a wide variety of functional groups, but moving to Hg^*/H_2 conditions, where H atoms are the abstractors, greatly extends the method. The method is likely to be useful for preparing unavailable starting materials for use in organic synthesis and for the commercial synthesis of high value compounds.

Acknowledgments.

I thank the U.S. Dept. of Energy for support of this work, Mark Burk for the original observation and Steve Brown, Cesar Muedas and Richard Ferguson for developing the method.

References.

[1] R.H. Crabtree, Chem. Rev., 85, (1985) 245.

[2] M.J. Burk and R.H. Crabtree, J. Amer Chem. Soc., 109 (1987) 8025.

[3] S.H. Brown and R.H. Crabtree, J. Amer Chem. Soc., 111 (1989), 2935 and 2946. C.A Muedas, R.R. Ferguson, S.H. Brown and R.H. Crabtree, J. Amer Chem. Soc., 113, (1991) 2233. S.H. Brown and R.H. Crabtree, US Pat. No., 4, 725,342, 1988. C.A Muedas, R.R. Ferguson, S.H. Brown and R.H. Crabtree, US Pat. Nos., 4, 874,488, 1989, and 5,104,503, 1992.

[4] G. Cario and C. Frank, Z. Physik, 1922, 11, 155.

[5] E.W.R. Steacie, "Atom and Free Radical Reactions", Reinhold, N.Y., 1954; R.J. Cvetanovic, Prog. React. Kinet., 2, (1963) 39.

[6] W.H. Breckenridge, C. Jouvet and B. Soep, J. Chem. Phys., 84, (1986) 1443.

[7] R.R. Ferguson and R.H. Crabtree, Nouv. J. Chim., 13, (1989) 647 and J. Org. Chem., 56, (1991) 550

Rh(III) Catalysts for Tail-to-Tail Dimerization of Methyl Acrylate

Maurice Brookhart, Elisabeth Hauptman and Sylviane Sabo-Etienne

Department of Chemistry, University of North Carolina
Chapel Hill, North Carolina 27599-3290, USA

Summary

The development of an efficient, highly selective Rh(III) catalyst system for the tail-to-tail dimerization of methyl acrylate is described. The catalytic cycle is entered by protonation of $Cp^*Rh(C_2H_4)_2$ to yield $Cp^*(C_2H_4)RhCH_2CH_2\text{-}\mu\text{-}H^+$, followed by reaction with methyl acrylate. The catalyst resting state has been generated by low-temperature protonation of $Cp^*Rh(CH_2CHCO_2Me)_2$ and identified as $Cp^*\overline{RhCH_2CH_2C(O)}OMe(\eta^2\text{-}CH_2CHCO_2Me)^+$. The turnover-limiting step is the C-C coupling reaction from the resting state. Additional low temperature NMR experiments provide a complete picture of the catalytic cycle. Investigation of Ir analogs has led to the isolation and X-ray structural characterization of $Cp^*\overline{IrCH_2CH_2C(O)}OMe(\eta^2\text{-}CH_2CHCO_2Me)^+$ thought to be isostructural with the Rh resting state.

1. Introduction

The catalytic tail-to-tail coupling of acrylate esters (Equation 1) is an alternative route to adipic acid from C_3 feedstocks and has been the subject of numerous investigations [1-3].

$$CH_2{=}CHCO_2R \longrightarrow RO_2CCH{=}CHCH_2CH_2CO_2R \;+\; RO_2CCH_2CH{=}CHCH_2CO_2R \qquad \text{Eq.1}$$

1a trans / **1b** cis (first product); **2a** trans / **2b** cis (second product)

Many of the early systems examined suffered from one or more drawbacks which included short catalyst lifetime, low turnover frequencies, formation of branched dimers and oligomers, and a requirement for high temperatures. After discovering that Rh(III) complexes of the type $C_5R_5(L)Rh(C_2H_4)(H)^+$ catalyze ethylene dimerization [4], and noting that such systems are compatible with functional groups, we were encouraged to investigate the possibility of using these species for acrylate dimerization. This contribution describes the development of such Rh(III) systems as long-lived catalysts for efficient and selective dimerization of acrylates [5]. A complete mechanistic study is also reported together with details of Ir analogs which model certain of the intermediates in the Rh(III) systems [6].

2. Catalyst Development

The initial complex selected for examination was $Cp^*(P(OMe)_3)Rh(C_2H_4)H^+$ **3** ($Cp^*= C_5Me_5$), our most efficient ethylene dimerization catalyst [4]. Treatment of **3** with methyl acrylate (MA) in CD_2Cl_2 at 25°C resulted in the formation of the chelate complex **4** followed by slow (less than one turnover/hour) tail-to-tail dimerization of MA. Complex **4** was synthesized independently by protonation of **5**. This work is summarized in Scheme I.

Scheme I

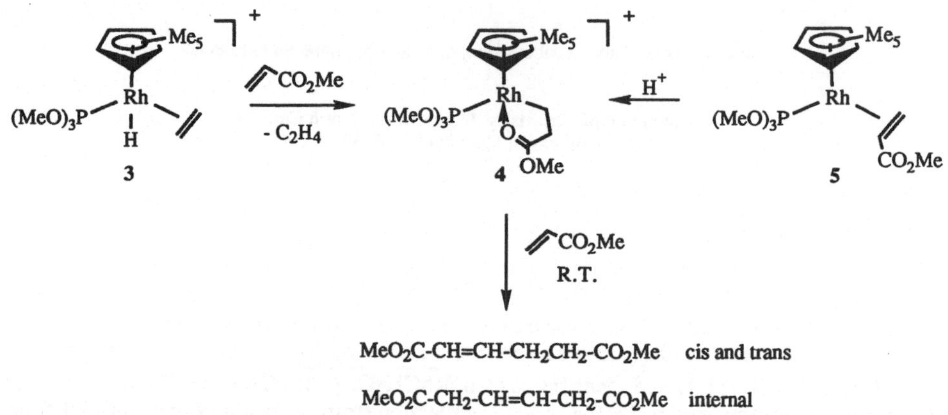

When $P(OMe)_3$ is replaced by ethylene, a much more reactive catalyst is generated. Thus, protonation of $Cp^*Rh(C_2H_4)_2$ leads to $Cp^*(C_2H_4)RhCH_2CH_2\text{-}\mu\text{-}H^+$, **6**, which upon exposure to MA gives exclusive tail-to-tail dimerization with an initial turnover rate of 6.6/min. The major dimer formed is the conjugated trans isomer **1a**; small amounts of **1b** are detected initially (ca. 10%). At the end of the reaction, traces of **2a** appear probably due to metal-catalyzed isomerization of **1a**, **1b**.

While the activity and selectivity of this system is high, the catalyst lifetime is limited. A maximum of 1200-1500 turnovers is achieved at 25°C. NMR investigations were conducted in an effort to discover the nature of the catalyst decomposition and possible ways to eliminate this problem. ^1H NMR monitoring of catalytic runs were carried out in CD_2Cl_2 where Cp* resonances were readily observed even in the presence of large excesses of MA. Treatment of **6** with MA (1600 equivalents) reveals that a new species, **7**, is immediately formed with a Cp* resonance at δ 1.6 ppm (the catalyst "resting state"). As dimerization proceeds, **7** slowly converts to a new species **8** (δ Cp*= 1.7 ppm) which represents the deactivated catalyst. Catalysis stops when **7** is totally depleted. When less than 100 equivalents of MA are used a new species **9** (δ Cp*= 1.53 ppm) appears at the end of the reaction. Addition of MA to this solution converts **9** to **7** and further dimerization ensues. This chemistry is summarized in Scheme II.

Scheme II

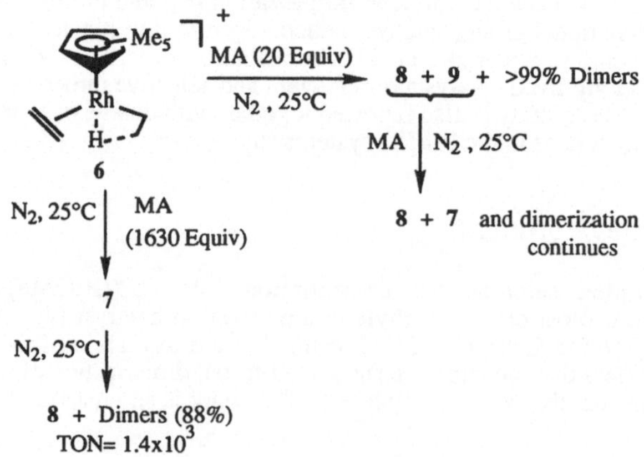

Complexes **8** and **9** have been isolated and identified by ^{1}H and ^{13}C NMR spectroscopy and X-ray analysis. Their structures are shown below.

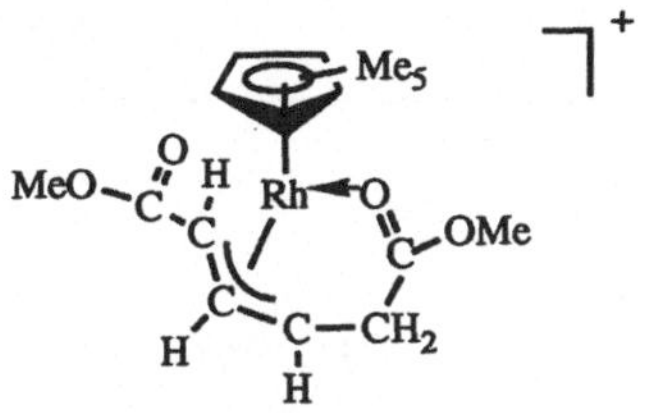

8 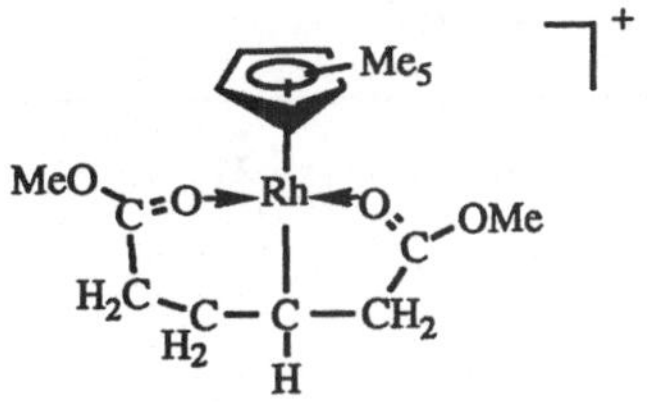 **9**

Several methods can be used to prepare these materials. Initially we observed that treatment of **6** with pure dimer gave a ca. 1:1 ratio of **8** and **9** which could be separated by fractional crystallization. Complex **8** can be isolated from solutions after catalysis ceases and **9** can be prepared from protonation of Cp*Rh(CH_2=$CHCO_2Me$)$_2$ (see below).
Noting that the deactivated catalyst **8** corresponds to loss of hydrogen, it occurred to us that it might be possible to reactivate the catalyst by exposure to H_2. This proved to be the case and an informative sequence of NMR experiments is shown in Scheme III.

Scheme III

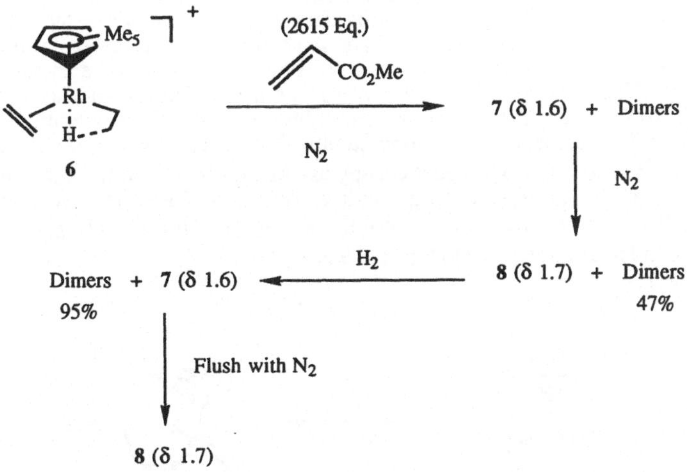

6 —(2615 Eq.) CH_2=CHCO_2Me, N_2→ **7** (δ 1.6) + Dimers

—N_2→ **8** (δ 1.7) + Dimers 47%

—H_2→ Dimers 95% + **7** (δ 1.6)

—Flush with N_2→ **8** (δ 1.7)

Since reactivation occurs under 1 atmosphere of H_2, it seemed additionally feasible to carry out the dimerization under H_2 and thereby achieve a continuously running long-lived catalyst system. Results summarized in Table I clearly show this is possible. Conducting the reaction at 60°C in neat MA under 1 atmosphere of H_2 results in complete conversion of 1.3×10^4 equivalents of MA with a 98% yield of tail-to-tail dimers and 2% yield of hydrogenated monomer, $CH_3CH_2CO_2Me$. Initial turnover rates correspond to 65 equiv. MA/min. The catalyst is still active at the end of this reaction so that 1.3×10^4 does not represent the maximum number of turnovers which can be achieved.

Table I. Catalytic dimerization of Methyl Acrylate (MA) using **6** under 1 atm of H_2

run	equiv of MA/[Rh]	T, °C	time, hrs	% conv	TON	ratio of MA dimers **1a: 1b: 2a**
1	6540	25	2	12		85: 15: 0
			5	34		85: 15: 0
			13	77		87: 13: 0
			20	97		92: 8: traces
			36	> 99	6.5 x 10^3	94: 5: 1
2	6540	60	1	65		89: 11: 0
			2	89		91: 9: traces
			3	94	6.2 x 10^3	94: 6: traces
	6540		3	47		94: 6: traces
	additional		4	55		94: 6: traces
	equiv. of MA		5	63		94: 6: traces
	added at this		6	70		94: 6: traces
	point		22	> 99	1.3 x 10^4	93: 6: 1

3. Mechanistic Studies

The results described above establish that a long-lived catalyst system can be operated under H_2 but at this point little is known concerning the mechanistic details of the catalytic cycle, in particular the structure of the resting state **7** and the nature of the turnover-limiting step. Reasoning that the resting state must involve two equivalents of acrylate, we prepared Cp*Rh(CH_2=$CHCO_2Me$)$_2$, **10**, and carried out a protonation of this complex at low temperature (-78°C) in the absence of monomer. Indeed, under these conditions a new species is formed with a Cp* resonance at 1.6 ppm identical to the resting state **7**. Characterization of this species by ^1H and ^{13}C NMR spectroscopy establishes its structure as the chelate complex shown in Scheme IV, in close analogy with **4**. Warming **7** results in formation of **9**. The formation of **9** can be envisioned as occuring via migratory insertion to give **11** followed by β-elimination and readdition to yield **9** (see Scheme IV).

Scheme IV

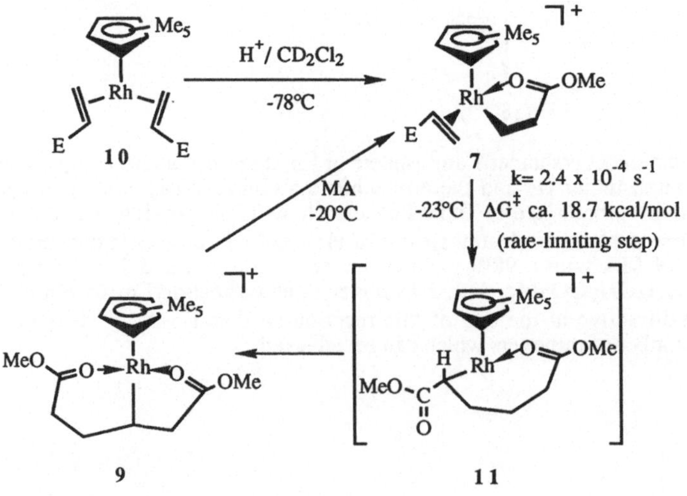

The first-order rate constant for isomerization of **7** to **9** was measured at -23°C and found to be 2.4×10^{-4} sec^{-1}, $\Delta G^{\ddagger} = 18.7$ kcal/mol. This free energy of activation is the same as that obtained earlier from the turnover frequency, 6.6/min at 25°C, $\Delta G^{\ddagger} = 19$ kcal/mol. The correspondence of these barriers clearly supports the identity of the resting state as **7** and the turnover-limiting step as the migratory insertion step. Further confirming the above chemistry, we have observed that treatment of **9** with MA at low temperatures (-20°C) results in regeneration of **7**.

From the results described thus far it is not clear whether **9** is an intermediate in the major catalytic cycle or whether an intermediate prior to formation of **9** is intercepted by MA to yield dimer(s) and regenerate **7**. Results of two experiments summarized in Scheme V clearly suggest that **9** is not in the major catalytic cycle. First, careful monitoring of the dimers formed when **9** is exposed to 25 equivalents of MA reveal that after ca. 3 turnovers the internal dimer **2a** constitutes ca. 40% of the dimers and after 3-4 turnovers 10-15% of **9** still remains. A second set of experiments (Scheme V) shows that treating pure **7** with MA (60 equivalents) and monitoring dimers formed at low temperatures results in no formation of internal dimer **2a**. The combination of these experiments clearly suggests that **9** upon reaction with MA releases almost exclusively internal dimer **2a**. Since only **1a** and **1b** are generated initially when the catalysis is initiated with **7**, **9** cannot be involved in the major catalytic cycle.

Scheme V

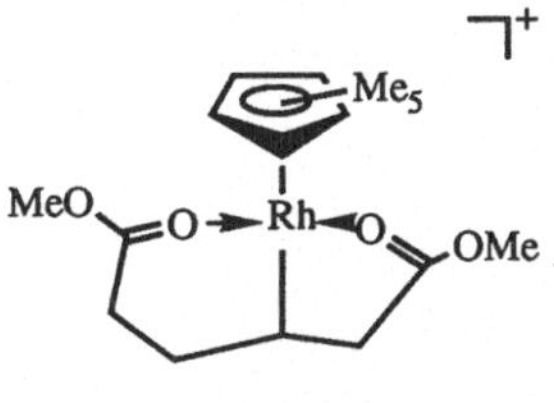

9 δ 1.53 — MA (25 eq.), -78°C → RT

TO number	internal	Dimers trans & cis	1.53 : 1.6
ca. 3	40%	60%	1 : 7
ca. 7	10%	90%	traces 1.53
100% conv.	5%	90% 5%	only 1.53

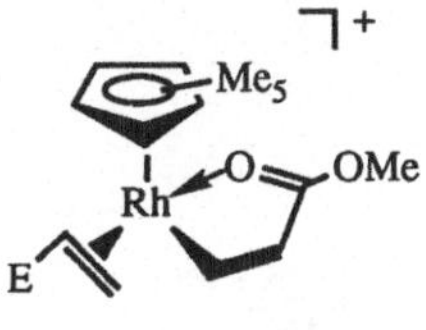

7 δ 1.6 — MA (60 eq.), -78°C → RT

TO number	internal	Dimers trans & cis
ca. 2.5	none	75% 25%
ca. 6	none	90% 10%
100% conv.	5%	90% 5%

The overall cycle can be described as in Scheme VI with complex **9** forming only at very low MA concentrations. While we depict the intermediate which is intercepted by MA as **11**, we have no structural evidence of this species and it may well be for example an oxoallyl species in which the carbonyl unit is coordinated to the metal to generate a formally 18-electron complex.

Scheme VI

CATALYTIC CYCLE

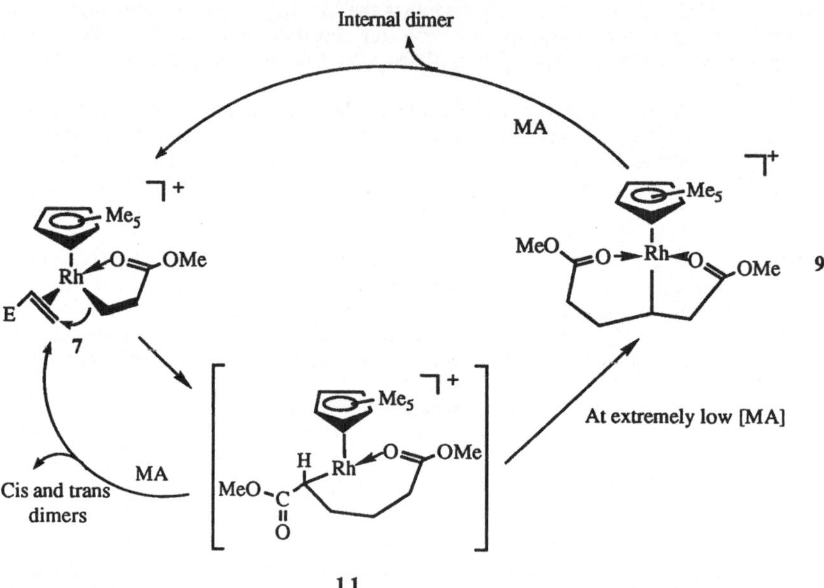

4. Iridium Analogs

To gain additional information regarding the intermediates in the catalytic cycle and the structure of the catalyst resting state, the iridium analogs were investigated. Protonation of $Cp^*Ir(C_2H_4)_2$, **12**, in the presence of MA leads sequentially to **13** and then to **14**, both of which can be isolated (Scheme VII).

Scheme VII

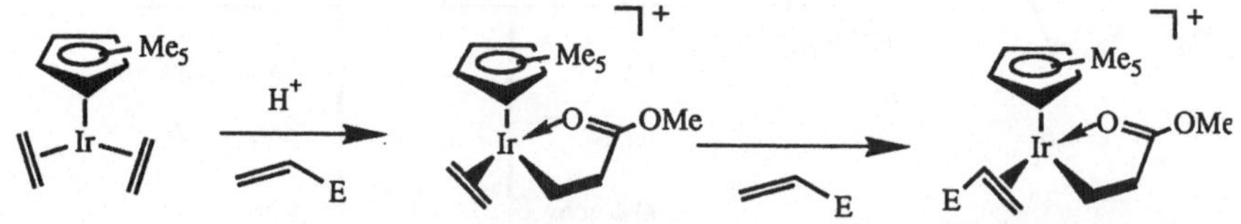

Me5 Ir **12** — H^+, E → Me5 Ir O OMe **13** (Fluxional, Ethylene Rotation) — E → Me5 Ir O OMe E **14a,b,c** (Fluxional, 3 isomers)

Variable temperature ^1H NMR spectroscopy reveals that **14** is a mixture of isomers **14a** (40%), **14b** (40%) and **14c** (20%). Isomers **14a** and **14b** are in rapid equilibrium via acrylate rotation. Cooling a hexane/diethyl ether solution of **14** results in crystallization of a single isomer which was identified as **14a** by X-ray crystallography (see Fig. 1).

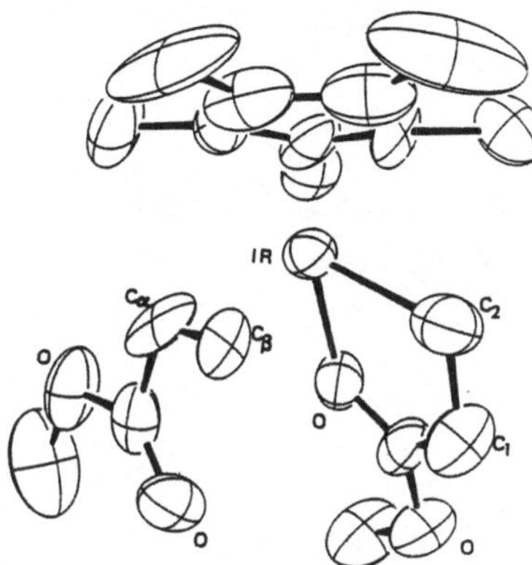

Figure 1. Ortep diagram of $Cp^*\overline{IrCH_2CH_2C(O)}OCH_3(\eta^2\text{-}CH_2CHCO_2CH_3)^+$ **14a**. The counterion $[3,5\text{-}(CF_3)_2C_6H_3]_4B^-$ as well as hydrogen atoms have been omitted.

Dissolution of pure **14a** in CD_2Cl_2 at -78°C results in generation of a 1:1 mixture of **14a** and **14b** as expected due to rapid rotation of the acrylate ligand. Upon warming to 23°C, the minor isomer **14c** grows in. This chemistry is summarized in Scheme VIII.

Scheme VIII

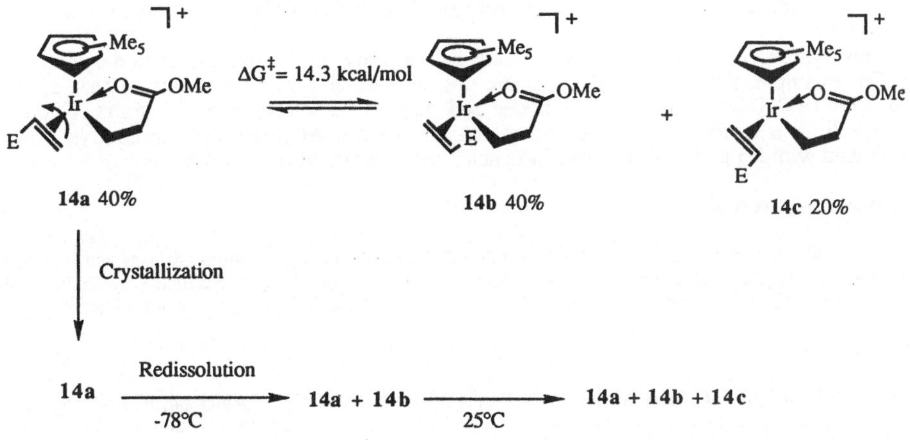

$\Delta G^{\ddagger}$= 14.3 kcal/mol

14a 40% ⇌ **14b** 40% + **14c** 20%

14a —Crystallization→ **14a** —Redissolution, -78°C→ **14a + 14b** —25°C→ **14a + 14b + 14c**

The orientation of the acrylate ligands in **14a** is that required for tail-to-tail coupling. Thus this complex is a model for the structure of the rhodium complex just prior to C-C coupling and is very likely isostructural with the rhodium resting state **7**.

5. Second Generation Catalysts

Several derivatives of **6** have been examined with the goal of increasing catalyst activity and lifetime. The most promising systems discovered to date are the trimethyl and heptamethyl indenyl systems [7] **15** and **16**. The activity of these systems is ca. 50% greater than that of the parent system, exhibiting turnover numbers of ca. 9-10/min compared to 6-7/min for **6** (see Scheme IX).

Scheme IX

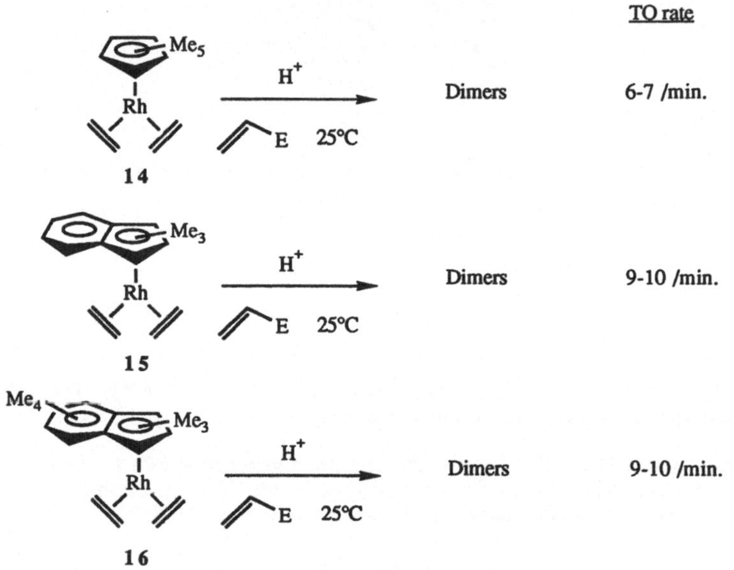

			TO rate
Me5 Rh **14**	H^+, E 25°C	Dimers	6-7 /min.
Me3 Rh **15**	H^+, E 25°C	Dimers	9-10 /min.
Me4 Me3 Rh **16**	H^+, E 25°C	Dimers	9-10 /min.

Preliminary mechanistic investigations reveal a catalytic cycle similar to that shown in Scheme VI.

The most noteworthy feature of these systems is their long catalytic lifetimes in the absence of H_2. For example, protonation of **15** in the presence of 60,000 equivalents of MA at 55°C results in 17,000 turnovers after 5 hours and 42,000 turnovers after 33 hours. Catalysis effectively ceases after 54,000 turnovers (ca. 68 hours). At this point the catalyst can be reactivated with H_2 and > 99% conversion achieved after an additional 6 hours.

Acknowledgments

We thank J. C. Calabrese (Du Pont) and P. S. White (UNC) for the X-ray structural analyses and P. J. Fagan (Du Pont) for helpful discussions. This work was supported by the National Science Foundation (CHE-8705534) and by E. I. Du Pont de Nemours & Co., Inc. We also thank Johnson Matthey for a loan of $RhCl_3$.

References

[1] Rh-based catalysts: (a) T. Alderson, U. S. Patent 3013066, 1961 (b) T. Alderson, E. L. Jenner, R. V. Lindsey, J. Am. Chem. Soc. 87 (1965) 5638 (c) W. A. Nugent, R. J. McKinney, J. Mol. Catal. 29 (1985) 65. (d) D. M. Singleton, U. S. Patent 4638084, 1987

[2] Pd- and Ni-based catalysts: (a) M. G. Barlow, M. J. Bryant, R. N. Haszeldine, A. G. Mackie, J. Organomet. Chem. 21 (1970) 215 (b) G. Oehme, H. Pracejus, Tetrahedron Lett. (1979) 343 (c) H. Pracejus, H. J. Krause, G. Oehme, Z. Chem. 20 (1980) 24 (d) W. A. Nugent, F. W. Hobbs, J. Org. Chem. 48 (1983) 5364 (e) W. A. Nugent, U. S. Patent 4451665, 1984 (f) see ref 1c (g) I. Tkatchenko, D. Neibecker, P. Grenouillet, FR Patent 2524341, 1983 (h) P. Grenouillet, D. Neibecker, I. Tkatchenko, FR Patent 2596390, 1987 (i) P. Grenouillet, D. Neibecker, I. Tkatchenko, Organometallics 3 (1984) 1130 (j) I. Guibert, D. Neibecker, I. Tkatchenko, J. Chem. Soc., Chem. Commun. (1989) 1850 (k) G. Wilke, K. Sperling, L. Stehling, Ger. Offen. DE 3336691, 1985

[3] Ru-based catalysts: (a) R. J. McKinney, U. S. Patent 4485256, 1986 (b) R. J. McKinney, M. C. Colton, Organometallics 5 (1986) 1080 (c) R. J. McKinney, Organometallics 5 (1986) 1752 (d) C. Y. Ren, W. C. Cheng, W. C. Chan, C. H. Yeung, C. P. Lau, J. Mol. Catal. 59 (1990) L1

[4] (a) M. Brookhart, D. M. Lincoln, J. Am. Chem. Soc. 110 (1988) 8719 (b) M. Brookhart, E. Hauptman, D. M. Lincoln, J. Am. Chem. Soc., submitted.

[5] M. Brookhart, S. Sabo-Etienne, J. Am. Chem. Soc. 113 (1991) 2777

[6] M. Brookhart, E. Hauptman, J. Am. Chem. Soc. 114 (1992) 4437

[7] (a) T. B. Marder, A. K. Kakkar, N. J. Taylor, J. C. Calabrese, W. A. Nugent, C. D. Roe, E. A. Connaway, J. Chem. Soc., Chem. Commun. (1989) 990 (b) D. O'Hare, J. C. Green, T. Marder, S. Collins, G. Stringer, A. K. Kakkar, N. Kaltsoyannis, A. Kuhn, R. Lewis, C. Mehnert, P. Scott, M. Kurmoo, S. Pugh, Organometallics 11 (1992) 48

Configurationally Stable and Configurationally Labile Chiral α-Substituted Organolithium Compounds in Stereoselective Transformations

Reinhard W. Hoffmann

Fachbereich Chemie der Philipps-Universität
Hans-Meerwein-Strasse, D-3550 Marburg

Summary

α-Bromo-alkyllithium compounds such as (**18**) have been generated by diastereoselective bromine/lithium exchange from the dibromoalkanes (**17**). On trapping with electrophiles, diastereomer ratios of up to 94:6 have been attained. The α-bromo-alkyllithium compounds have been found to be configurationally stable at -110°C. In contrast, the α-phenylseleno-alkyllithium compounds (**24**) epimerise rapidly at -78°C. Nevertheless diastereomer ratios of up to 93:7 may be realized upon quenching with electrophiles. The configurational stability of these and related α-heterosubstituted organolithium have been rapidly assessed by a test based on kinetic resolution.

1 Introduction

Addition of organolithium compounds to aldehydes, ketones or epoxides is one of the standard transformations in stereoselective synthesis. Much can be gained, if chiral α-heterosubstituted organolithium compounds are used as chiral building blocks in such transformations. This is illustrated by an outstanding example from D. Hoppe's group [1].

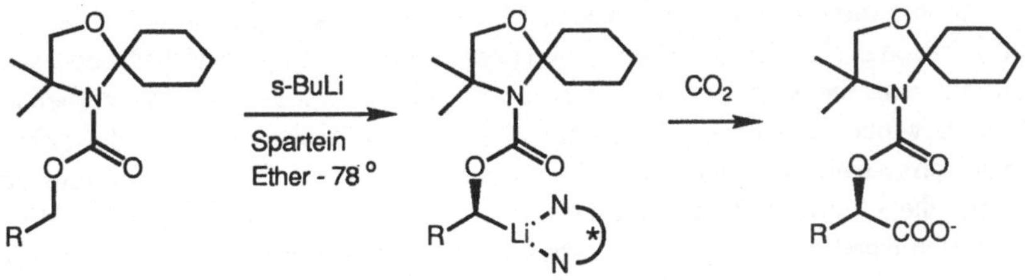

s-BuLi

Spartein
Ether - 78 °

CO_2

Scheme 1

Before such a strategy is generally applicable to other chiral organolithium compounds (**1**), positive information on the following two questions is required:

- Is there a temperature range and a solvent system, in which the organolithium compounds (**1**) cannot only be generated and reacted, but in which they are configurationally stable ?
- Are there ways to generate such organolithium compounds as enantiomerically pure entities?

1

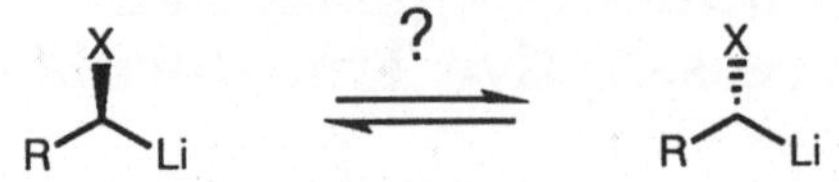

?

X = Br, SePh, SPh, OCH_2OCH_3, $OCON(iPr)_2$

Scheme 2

Therefore at the beginning of the development of this kind of chemistry the configurational stability of the entities (**1**) had to be evaluated. Thus, W.C. Still [2] demonstrated in a pioneering study that α-alkoxyalkyllithium compounds remain configurationally stable for at least 15 min at -30°C in THF:

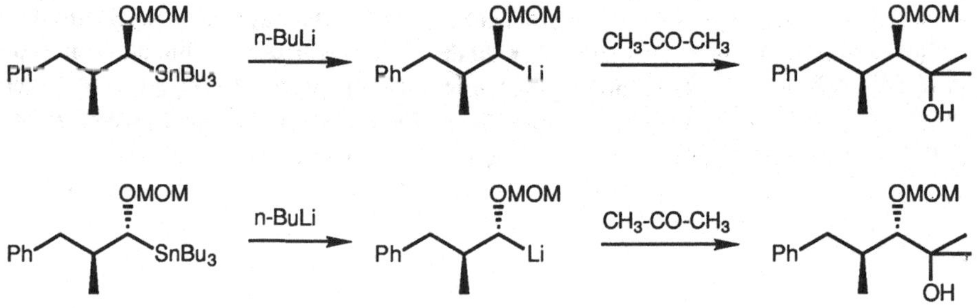

n-BuLi

CH_3-CO-CH_3

Scheme 3

In due course the configurational stability of sulfur- [3,4,5] and seleno- [5,6,7] substituted alkyllithium compounds, as well as of nitrogen-substituted derivatives [8,9] (**1**) have been evaluated by the same type of experiments. This way, a considerable body of knowledge on the configurational stability of α-substituted organolithium compounds had been accumulated. Our interest focussed on the α-bromo-alkyllithium compounds, the carbenoids, the chemistry of which goes back to the seminal studies of Koebrich [10] and Seyferth [11]. Their work on α-halo-cyclopropyllithium compounds, indicated (low) temperature ranges in which both the chloro- [12] and bromo-derivatives [13] remain configurationally stable. The well established chemistry of the α-bromo-alkyllithium compounds [14] encouraged us to explore their synthetic potential in stereoselective transformations. Again the outset of our study information on the configurational stability of these entities had to be sought.

2 The Configurational Stability of Chiral α-Heterosubstituted Alkyllithium Reagents

When W.C. Still studied the configurational stability of α-alkoxyalkyllithium compounds, [2] he used as outlined above a second stereocenter in the molecule as a reference point, relative to which the configurational stability of the lithium bearing stereocenter was evaluated. Both diastereomeric educts had to be prepared and studied to attain definitive results. Recently, Prof. Hoppe's and our group proposed [15] a test [16] which provides qualitative information on the configurational stability of a chiral organometallic reagent. The test demonstrates whether the organometallic reagent reacts more slowly or more rapidly with a chiral electrophile than it racemizes. The advantage of this test is that it may be carried out on the racemic organometallic species.

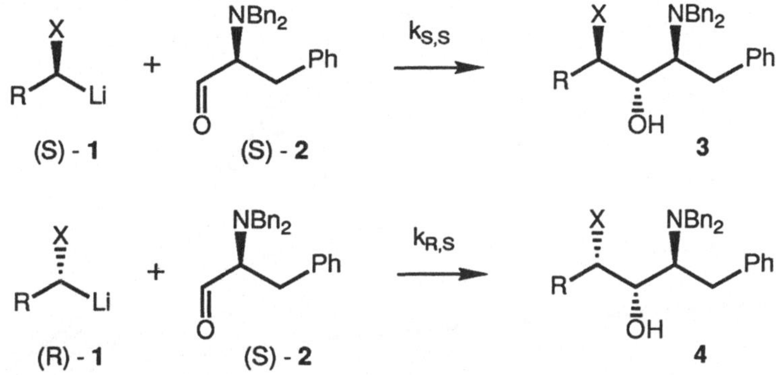

(S) - **1** + (S) - **2** $\xrightarrow{k_{S,S}}$ **3**

(R) - **1** + (S) - **2** $\xrightarrow{k_{R,S}}$ **4**

Scheme 4

The test is based on kinetic resolution, which occurs when two racemates react with each other: This is illustrated in the reaction of the organolithium compound (1) with the chiral aldehyde (2) [17] as the electrophile. For instance, the (S)-enantiomer of the organolithium compound (1) reacts with the (S)-aldehyde (2) to give the product (3); in turn the (R)-enantiomer of (1) leads on reaction with the (S)-aldehyde (2) to the diastereomeric product (4). Since these are two different reactions, their rate constance k_{SS} and k_{RS} are likely to be different. This difference, i.e. the presence of kinetic resolution, constitutes the prerequisite for the test. The ratio k_{SS}/k_{RS} can be determined from a reaction of the racemic organolithium compound (1) with the racemic aldehyde (2) (electrophile), giving the two diastereomeric racemic products (3) and (4) in a ratio p. This product ratio p is equal to k_{SS}/k_{RS}. For the subsequent test, this ratio should be $\neq 1$; the optimal range for p is between 1.5 and 3.0. In a second experiment, the enantiomerically pure aldehyde (2) is allowed to react with the racemic organolithium compound. If (S)-(1) and (R)-(1) interconvert rapidly, i.e. more rapidly than they add to the aldehyde (2), the ratio q of the products (3) and (4) is defined by $K \cdot k_{SS}/k_{RS}$ in which K is the equilibrium constant between the two enantiomers of (1). Since $K = 1$, the product ratio q should be equal k_{SS}/k_{RS}, as in the first experiment. Therefore, if q is found to be equal to p, the organolithium compound is configurationally labile on the time scale set by the rate of reaction of (1) with the aldehyde (2) (the electrophile).

If (S)-(1) and (R)-(1) do not interconvert, or at least more slowly than they add to the aldehyde (2), (S)-(1) should be completely converted into (3), and (R)-(1) into (4). Since there are equal amounts of (S)-(1) and (R)-(1), in the racemate the two products (3) and (4) should be formed in equal amounts, i.e. q should be = 1 after complete conversion of (1) into the products. The evolution of the diastereomer ratio q over the %-conversion is illustrated in fig. 1, for various values of the kinetic resolution $k_{SS}/k_{RS} = s$.

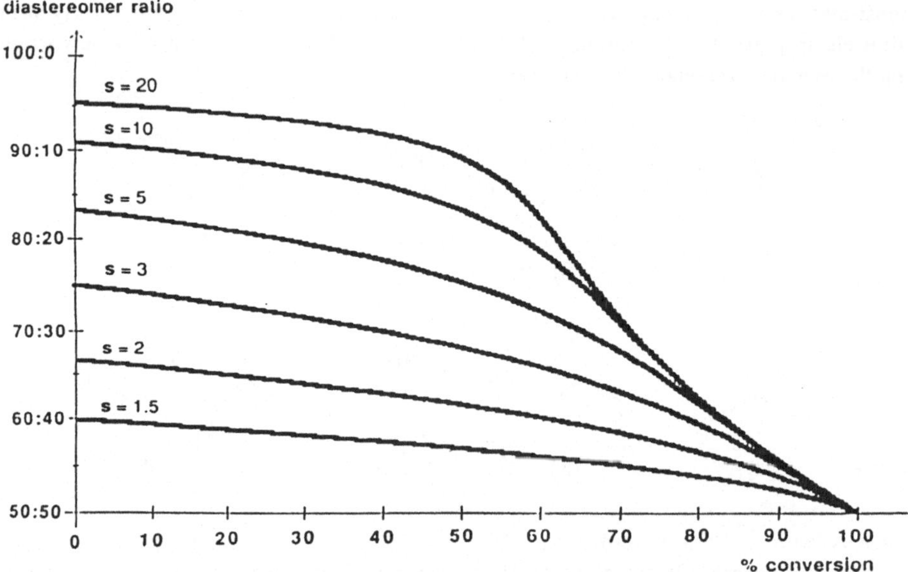

Fig. 1 Evolution of the diastereomer ratio q as a function of % conversion in the reaction of a racemate with an enantiomerically pure reagent. S is the kinetic resolution.

At low conversion, the product ratio is essentially that defined by the kinetic resolution, i.e. $q \approx p$. Once, the conversion is larger than 50% the product ratio approaches the value 1, which is reached eventually at 100% conversion. As soon as the conversion in experiment 2 is > 50%, any analytically significant difference between the product ratio q in experiment 2 and p in experiment 1 signals, that the organolithium species (1) is configurationally stable on the time scale set by the rate of reaction of (1) with the electrophile. The operational scheme is summarized below:

Test on the Configurational Stability Based on Kinetic Resolution

Expt. 1 rac - Substrate + **rac** - Electrophile → Diastereomers A + B, Ratio = **p**

if $p \approx 1$,	Choose other Electrophile
if $p \neq 1$, (best 1.5 -3.0)	Go to Expt. 2

Expt. 2 rac - Substrate + **e.p.** - Electrophile → Diastereomers A + B, Ratio = **q**

if $q = p$,	Substrate is Config. LABILE
if $q \neq p$, (usually $q \rightarrow 1$)	Substrate is Config. STABLE

We have applied this test to evaluate the configurational stability of the simple α-bromo-alkyllithium compound (5) [18].

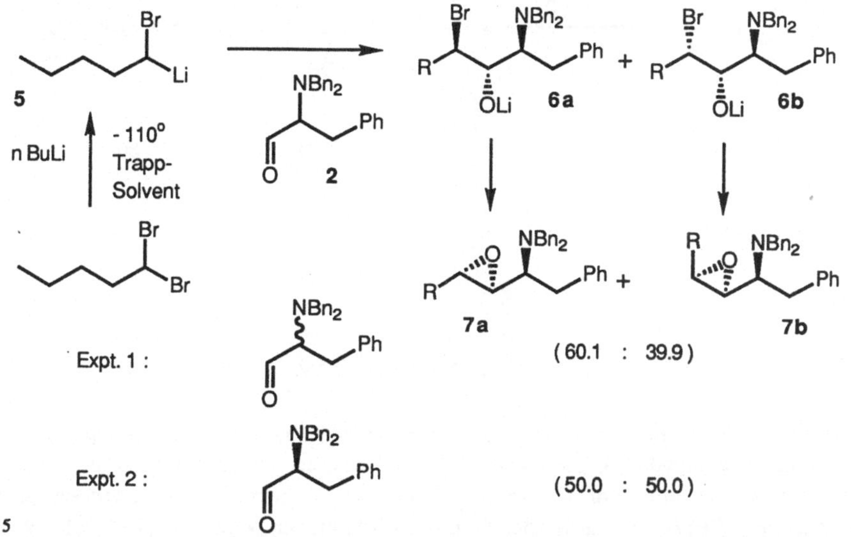

Br
Li
5
n BuLi
- 110°
Trapp-
Solvent
Br
Br
NBn2
Ph
O 2
Br NBn2
R Ph
OLi 6a
+
Br NBn2
R Ph
OLi 6b
NBn2
R O Ph
7a
+
R NBn2
O Ph
7b
Expt. 1 :
NBn2
Ph
O
(60.1 : 39.9)
Expt. 2 :
NBn2
Ph
O
(50.0 : 50.0)

Scheme 5

After reaction of (5) with the aldehyde (2) the primary adducts (6) cyclized to the epoxides (7). It is not necessary to assign the stereostructures to the two adducts (7); it is simply their ratio that is diagnostic. The difference between p from experiment 1 and q from experiment 2 revealed the "configurational stability" of the α-bromo-alkyllithium compounds of interest. Likewise, the configurational stability of the α-phenylseleno- (8) [7] and the α-phenyl-thio-alkyllithium compounds (9) [19] were assessed by this test:

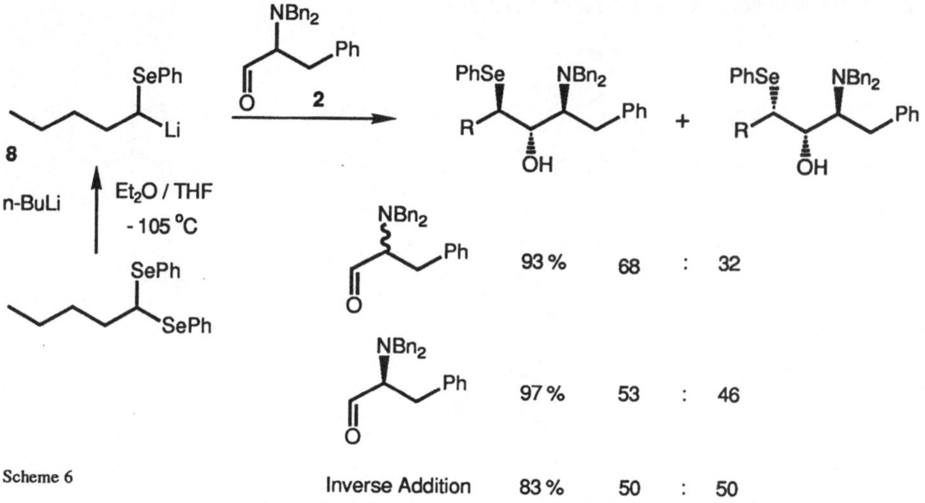

NBn2
Ph
O 2
SePh
Li
8
n-BuLi
Et_2O / THF
- 105 °C
SePh
SePh
PhSe NBn2
R Ph
OH
+
PhSe NBn2
R Ph
OH

NBn2 (racemic) Ph, O	93 %	68	:	32
NBn2 Ph, O	97 %	53	:	46
Inverse Addition	83 %	50	:	50

Scheme 6

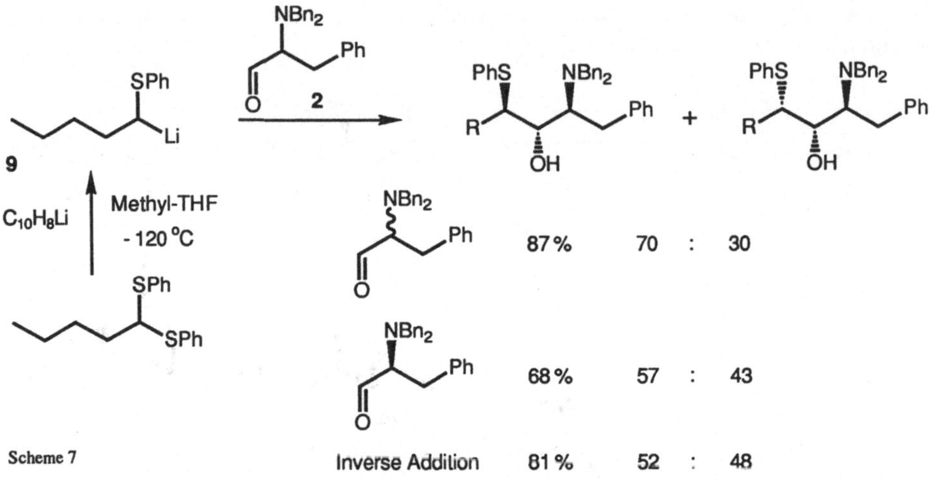

Scheme 7

It should be added, that it is necessary in experiment 2 (but not in experiment 1) to add the organolithium compound to the electrophile, in order to maintain always an excess of the electrophile. The effect of the order of addition is seen in the reactions of the α-phenylseleno- (**8**) [7] and the α-phenylthio-alkyllithium compound (**9**) [19] with the aldehyde (**2**). The differences between the normal and inverse addition modes in experiment 2 indicate, that partial equilibration between the enantiomers of (**8**) or of (**9**) competes with the addition process, as long as the stationary concentration of the aldehyde (**2**) is low, i.e. under such conditions the rate of trapping of (**8**) by (**2**) is no longer faster, than the racemisation of (**8**). Thus, the high tendency of these organolithium compounds for stereorandomisation, which was noticed from other studies [4,5,6,7], is substantiated by these experiments. The configurational stability of such entities is increased in going from the lithium to the corresponding Grignard reagent (**10**) [19]:

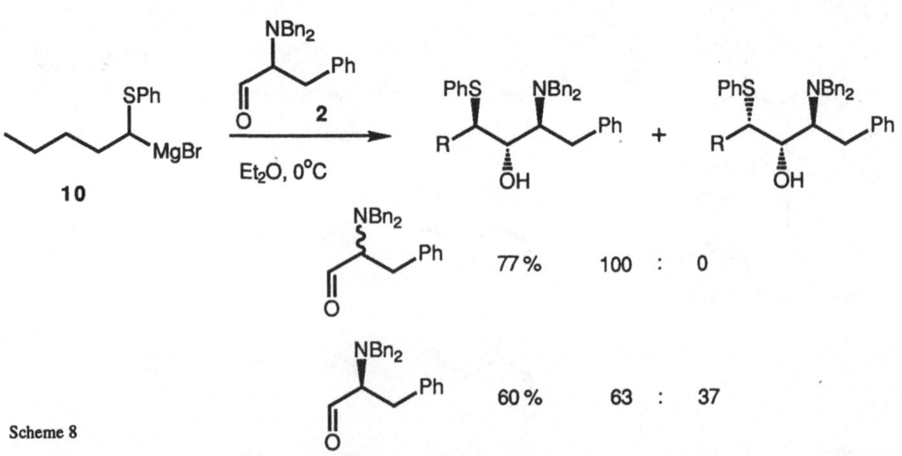

Scheme 8

In this context it is of interest, whether added lithium halides, or the addition of complexing agents for the lithium cation have a significant influence on the rate of the enantiomer equilibration. W.H. Pearson [8] noted that addition of TMEDA to (**11a**) resulted in a very fast epimerization at the lithium bearing stereocenter. Likewise J.M. Chong described that a addition of HMPT leads to a rapid racemisation of (**11b**) [20]. This contrasts the finding of P. Beak [9] that TMEDA enhanced the configurational stability of (**12**).

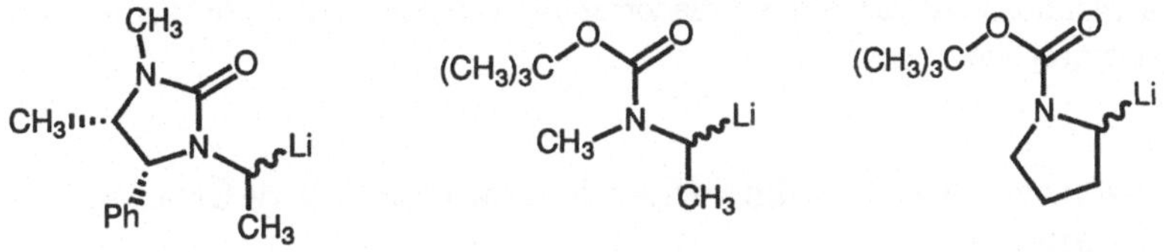

Scheme 9 **11a** **11b** **12**

Could relevant information also be gained by using the above mentioned test? To learn more about such systems, we investigated the reaction of the α-bromo-alkyllithium compound (**5**) with 2-benzyloxy-propionaldehyde (**13**):

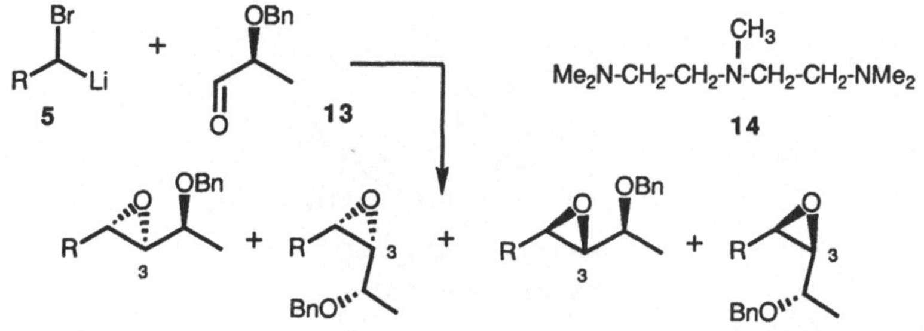

5 **13** **14**

	15a	**15b**	**15c**	**15d**	
rac-**13**	18.8	30.7	20.3	29.2	p = 0.98
rac-**13** + 2 equiv. PMDTA	71.0	5.8	6.2	16.9	p = 3.3
(S)-**13** + 2 equiv. PMDTA	53.8	3.4	9.9	32.9	q = 1.3

Scheme 10

Reaction of (**5**) with the racemic aldehyde **13** led to four diastereomeric epoxides (**15**), rather than the two adducts (**7**) obtained with (**2**). This is a consequence of the low asymmetric induction of the aldehyde (**13**) on the formation of the stereocenter at C-3. In this case, the relative configuration of the products (**15**) had to be assigned, in order to find out, whether kinetic resolution, i.e. the ratio (**15a** + **15b**): (**15c** + **15d**) is sufficiently different from 1.0. The assignment of the stereostructures [21] made it clear that the kinetic resolution in this particular experiment was unfortunately too small (p ≈ 1) to allow the application of our test. This situation changed however, when the experiment 1 was run in the presence of pentamethyl-diethylene-triamine (PMDTA) (**14**). Now, the product ratio revealed [22] a sufficient kinetic resolution of 77:23. This change, by the way, indicates that the complexed organolithium compound has a higher reactivity than the uncomplexed one, and that (predominantly?) the former adds to the aldehyde. The experiment 2 with the enantiomerically pure aldehyde (**13**) run in the presence of PMDTA now led to a product ratio (**15a** + **15b**):(**15c** + **15e**) = q = 57:43 at 90% conversion. The difference in the values of q and p signals, that the reactive entity is trapped faster by the aldehyde (**13**) than it racemices. Apparently, the configurational stability of the organolithium compound (**5**) is not

Thus, these examples demonstrated how the described test provides rapid qualitative information on the configurational stability of organometallic reagents by comparing the rate of enantiomerization of the organometallic compound with the rate of its addition to a chiral electrophile.

3 Generation of Configurationally Homogenous α-Bromoalkyl-lithium Compounds

While configurational stability is one of the prerequisites for utilization of chiral α-hetero-substituted organolithium compounds (1) in synthesis, the other prerequisite are ways to generate them as configurationally homogeneous entities. In this context we explored the selective bromine/lithium exchange of the diastereotopic bromine atoms in compounds of the type (17).

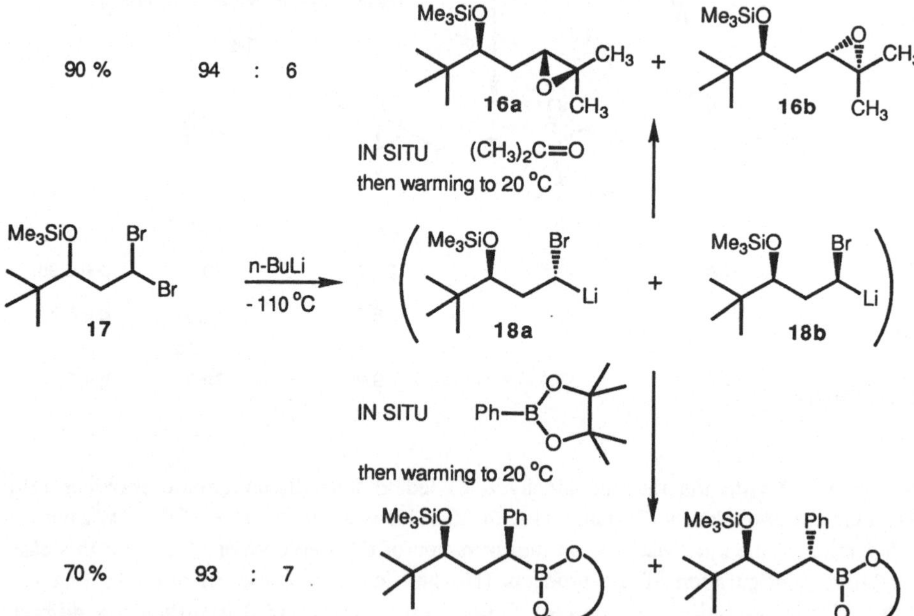

90 % 94 : 6

Me_3SiO **16a** CH_3 CH_3 + Me_3SiO **16b** CH_3 CH_3

IN SITU $(CH_3)_2C{=}O$
then warming to 20 °C

Me_3SiO Br Br **17** — n-BuLi, - 110 °C → (Me_3SiO Br Li **18a** + Me_3SiO Br Li **18b**)

IN SITU Ph—B(pinacolato)
then warming to 20 °C

70 % 93 : 7

Me_3SiO Ph B **19a** + Me_3SiO Ph B **19b**

Scheme 11

While nature makes extensive use of the differentiation of diastereotopic groups, surprisingly few reports exist on the differentiation of diastereotopic groups as a means of stereoselective synthesis. We were pleasantly surprised that our first experiment on the reaction of (17) with n-butyllithium proceeded with > 90 % selectivity [23]: The intermediate lithium species (18) could be generated in the presence of either ketones or phenylboronates. In situ quenching by acetone led to the two epoxides (16) in a straightforward reaction. The in situ trapping by phenylboronate is more complex and involves formation of a boron-ate-complex which rearranges stereospecifically [24] forming a phenyl-carbon bond. The resulting alkylboronates are of general preparative interest, because the carbon-boron bond may chemistry; i.e. it can be oxidized generating an alcohol function in a stereospecific manner.

The high stereoselectivity in the overall process (17) → (16) or (17) → (19) did not, however, establish whether the intermediate α-bromo-organolithium-compounds (3) had equilibrated prior to being trapped by the electrophile. When the addition of the electrophile was delayed, in a study carried out with the dibromo compound (20), it was found that the ratio of the diastereomeric epoxides (22) remained constant [25] (cf. the left diagram in scheme 12).

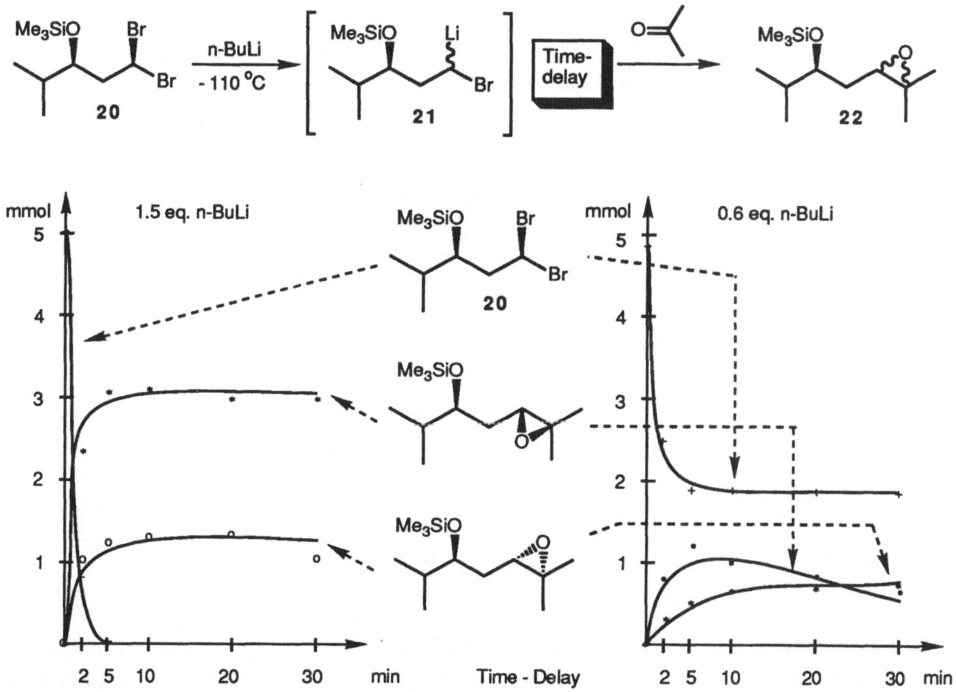

Scheme 12

This could mean that the ratio of the intermediary α-bromo-alkyllithium compounds is under kinetic control or that the ratio of (21) is accidentally close to the thermodynamic ratio. It became clear that the former situation holds, once it was found that these intermediates may be equilibrated by the action of an excess of the dibromo compound (20) [25], a reaction that proceeds via a reversible bromine/lithium-exchange between (21) and (20), as was found initially by Seyferth [11,26], (cf. the right diagram in scheme 12).

It is thus advisable to generate the α-bromo-alkyllithium compounds (18) or (21) from the dibromo compounds (17) and (20) in the presence of the electrophile (in situ trapping) to achieve high diastereoselectivity [23]. Otherwise the selectivity is compromised by partial equilibration of the α-bromo-alkyllithium compounds during their generation involving the dibromo compounds (12) or (15) still present.

The selective exchange of one of the two diastereotopic bromine atoms by lithium in compounds of the type (**17**) constitutes thus a convenient way to generate α-bromo-alkyllithium compounds of defined configuration. While the selectivity is particularly high with (**17**), moderate to good diastereoselectivities have been obtained with related systems as shown below [23,27,28]. By determining the relative configuration of the products such as (**16**), (**19**) or (**22**) it became apparent which of the two bromine atoms, the *ul*- or the *lk*-bromine atom is exchanged in preference. The reasons, for which a particular bromine atom is preferentially exchanged are still a matter of conjecture. A conformational analysis revealed [29] that the ground state conformer population of e.g. (**20**) is heavily biassed (>90% at -110°C) in favour of the conformation shown below. If this is also the reactive conformation it is the most readily accessible *ul*-bromine atom which is exchanged in preference, as this has no gauche interactions with other groups in the molecule.

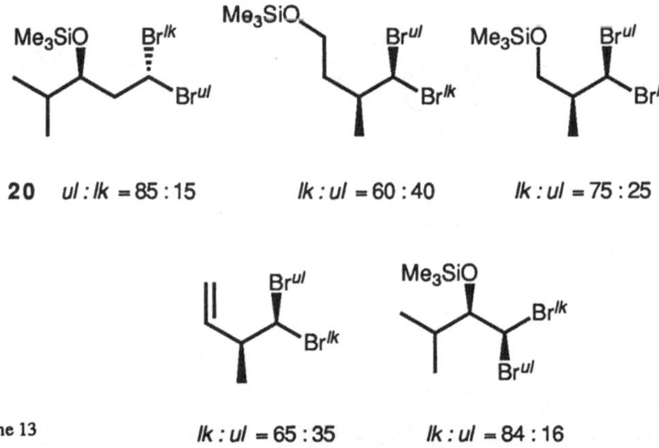

Scheme 13

The configurational stability of the α-bromo-alkyllithium compound was a key to the stereoselective generation of the products such as (**16**) or (**19**). Nevertheless, configurational stability is not a necessary prerequisite for attaining high diastereoselectivity. That high selectivities may be realized even with configurationally labile α-substituted organolithium compounds is demonstrated in the following section.

4 Generation and Utilization of Configurationally Labile α-Phenylseleno-Alkyllithium Compounds

It was known from the investigations of Reich [5] and Krief [6], as well as from our studies [7] mentioned before, that the enantiome interconversion of α-phenylseleno-alkyllithium compounds is a facile process. This is supported by a study of (**24**) which was generated by seleno/lithium exchange from the seleno acetal (**23**) corresponding to (**20**). It was found that (**24**) epimerized rapidly before the electrophiles could be added. The generation of (**24**) in the presence of the electrophiles (in situ trapping) was not possible as a consequence of the slow seleno/lithium exchange. Nevertheless, diastereoselectivities in excess of 85%

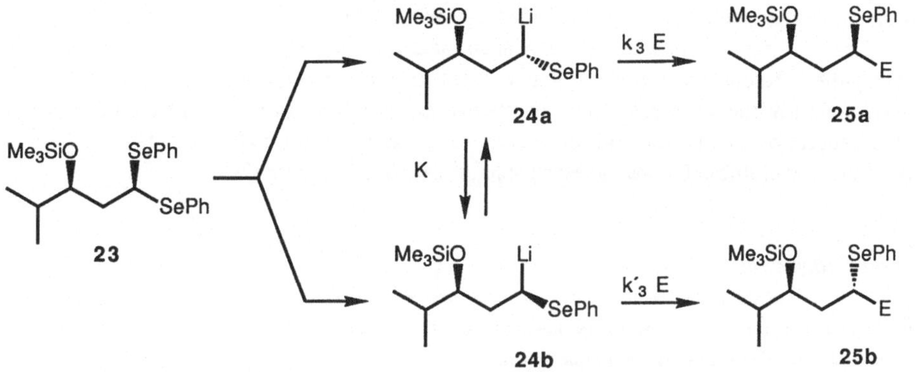

Me_3SiO Li Me_3SiO SePh

k_3 E

SePh E

24a 25a

Me_3SiO SePh

K

SePh

23

Me_3SiO Li Me_3SiO SePh

k'_3 E

SePh E

24b 25b

E =	a : b	Cyclohexanone	90 : 10 71 %
i-Butyraldehyde	93 : 7 78 %	Cyclohexenone	86 : 14 89 %
Acetone	91 : 9 93 %	Cyclopentanone	85 : 15 58 %

Scheme 14

The diastereomer ratio of the products **(25)** varies only to a small extent. This suggests that the diastereomer ratio reflects the equilibrium ratio **(24a)** $\rightleftharpoons$ **(24b)**, which lies substantially on the side of **(24a)**, and that the trapping of **(24)** by the electrophile is (at least slightly) faster than the rate of epimerization of **(24)**. Only when the rate of trapping becomes slower than the rate of epimerization of **(24)** is there an element of kinetic diastereoselection in the trapping step. This holds e.g. for the reverse Brook rearrangement of **(24)** to give the silyl compounds **(26)** [30], a reaction that occurs on warming etheral solutions of **(24)** to -40°C, or on generating **(24)** at -78°C in THF.

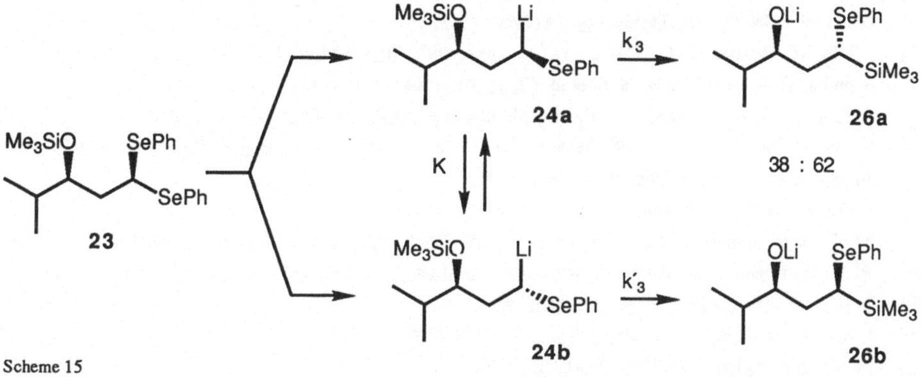

Me_3SiO Li OLi SePh

k_3

SePh $SiMe_3$

24a 26a

Me_3SiO SePh

K 38 : 62

SePh

23

Me_3SiO Li OLi SePh

k'_3

SePh $SiMe_3$

24b 26b

Scheme 15

In the reverse Brook rearrangement the diastereomer ratio of the products **(26)** no longer reflects the equilibrium ratio of the α-phenylseleno organolithium compounds **(24)**. It is rather determined by the equilibrium constant K and the two rate constant k_3 in a composite manner.

These studies presented here on the α-heterosubstituted organolithium compounds (**1**) shed some light on the competition between enantiomerization (epimerization) and trapping by electrophiles. Knowledge on the relative rates of these processes is essential when one wants to control the stereoselectivity in the overall transformations, be it by kinetic control in the generation of the organolithium compounds as in the formation of (**18**), or be it by letting them equilibrate before the electrophiles are added, as with (**24**).

Acknowledgement:

These studies were supported by the Deutsche Forschungsgemeinschaft (SFB 260) and the Fonds der Chemischen Industrie. I thank all my coworkers, the names of which are given in the references, for their perseverance in exploring this field of organometallic chemistry.

References

[1] D. Hoppe, F. Hintze, P. Tebben, Angew. Chem. **102** (1990) 1457; Angew. Chem. Int. Ed. Engl. **29** (1990), 1422.

[2] W. C. Still, C. Sreekumar, J. Am. Chem. Soc. **102** (1980) 1201.

[3] R. H. Ritter, T. Cohen, J. Am. Chem. Soc. **108** (1986) 3718.

[4] [4a)] P. G. McDougal, B. D. Condon, M. D. Laffosse, jr., A. M. Lauro, D. VanDerveer, Tetrahedron Lett. **29** (1988) 2547. - [4b)] G. P. Lutz, A. P. Wallin, S. T. Kerrick, P. Beak, J. Org. Chem. **56** (1991) 4938.

[5] H. J. Reich, M. D. Bowe, J. Am. Chem. Soc. **112** (1990) 8994.

[6] A. Krief, G. Evrard, E. Badaoui, V. De Beys, R. Dieden, Tetrahedron Lett. **30** (1989) 5635.

[7] R. W. Hoffmann, M. Julius, K. Oltmann, Tetrahedron Lett. **31** (1990) 7419.

[8] W. H. Pearson, A. C. Lindbeck, J. Am. Chem. Soc. **113** (1991) 8546.

[9] S. T. Kerrick, P. Beak, J. Am. Chem. Soc. **113** (1991) 9708.

[10] G. Köbrich, Angew. Chem. **79** (1967) 15; Angew. Chem. Int. Ed. Engl. **6** (1967), 41.

[11] D. Seyferth, R. L. Lambert, jr., J. Organomet. Chem. **55** (1973) C 53.

[12] G. Köbrich, W. Goyert, Tetrahedron **24** (1968) 4327.

[13] [13a)] P. M. Warner, S.-C. Chang, N. J. Koszewski, Tetra hedron Lett. **26** (1985) 5371. - [13b)] A. Schmidt, G. Köbrich, R. W. Hoffmann, Chem. Ber. **124** (1991) 1253.

[14] J. Villieras, B. Kirschleger, R. Tarhouni, M. Rambaud, Bull. Soc. Chim. Fr. **1986** 470.

[15] R. W. Hoffmann, J. Lanz, R. Metternich, G. Tarara, D. Hoppe, Angew. Chem. **99** (1987) 1196; Angew. Chem., Int. Ed. Engl. **26** (1987), 1145.

[16] R. Hirsch, R. W. Hoffmann, Chem. Ber. **125** (1992) 975 .

[17] M. T. Reetz, Angew. Chem. **103** (1991) 1559; Angew. Chem. Int. Ed. Engl. **30** (1991), 1531.

[18] R. W. Hoffmann, T. Ruhland, M. Bewersdorf, J. Chem. Soc., Chem. Commun. **1991** 195.

[19] F. Chemla, unpublished results, Marburg 1991.

[20] J. M. Chong, S. B. Park, J. Org. Chem. **57** (1992) 2220.

[21] M. Krüger, Diplomarbeit Univ. Marburg, 1988.

[22] T. Ruhland, Diplomarbeit Univ. Marburg, 1989.

[23] R. W. Hoffmann, M. Bewersdorf, M. Krüger, W. Mikolaiski, R. Stürmer, Chem. Ber. **124** (1991) 1243.

[24] R. Tripathy, R. W. Franck, K. D. Onan, J. Am. Chem. Soc. **110** (1988) 3257.

[25] R. W. Hoffmann, M. Bewersdorf, Chem. Ber. **124** (1991) 1259.

[26] D. Seyferth, R. L. Lambert, jr., M. Massol, J. Organomet. Chem. **88** (1975) 255.

[27] R. W. Hoffmann, M. Julius, Liebigs Ann. Chem. **1991** 811.

[28] W. Mikolaiski, Dissertation Univ. Marburg, 1990.

[29] R. W. Hoffmann, K. Brumm, M. Bewersdorf, W. Mikolaiski, A. Kusche, Chem. Ber. **1992** submitted.

[30] R. W. Hoffmann, M. Bewersdorf, Tetrahedron Lett. **31** (1990) 67.

Synthetic Aspects of the Metal-Mediated Cyclooligomerization of Phosphaalkynes [1]

Manfred Regitz

(Based on investigations in cooperation with Bernhard Breit, Manfred Birkel, Thomas Wettling, Bernhard Geißler, Uwe Bergsträßer, Heinrich Heydt, Stefan Barth, Charles Crittel, Paul Binger, Kenneth Laali, and Peter Stang)

Fachbereich Chemie der Universität Kaiserslautern, Erwin-Schrödinger-Strasse, D-6750 Kaiserslautern, Federal Republic of Germany

Summary

Although the thermal cyclooligomerization of the phosphaalkyne **10** (≡ **4**; R = *t*Bu) proceeds non-selectively to furnish a mixture of **11**, **12**, and **13**, the tetraphosphacubane **11** can be obtained in high yield from the reaction of the zirconium complex **15** with hexachloroethane. This reaction provides the starting point for numerous functionalization reactions at the phosphorus atoms of the pentacyclic system (⟶ **27-38**, **40**, and **42-46**). In the presence of aluminium trihalides the phosphaalkynes **4** (R = *t*Bu, 1-Ad) undergo spirocyclization with incorporation of the Lewis acid to furnish the 1,3-diphosphete betaines **50**. Removal of the Lewis acid moiety by treatment with DMSO induces a rearrangement to the Dewar phosphabenzenes **52** and **53** which, in turn, can be trapped by homo-Diels-Alder reactions with the phosphaalkyne **10** (⟶ **54**, **55**). Isomerization reactions among the phosphaalkyne cyclotetramers **54**, **55**, **13**, and **57** are discussed; they are in accord with the calculated heats of formation for the respective, unsubstituted molecules.

1 Introduction

The thermally unstable phosphaacetylene, HC≡P (**63**), was generated for the first time in 1961 by means of an electric discharge between carbon electrodes in a PH_3 atmosphere [2]. A further twenty years passed before the synthesis of tert-butylphosphaacetylene (**10** ≡ **4**, R = *t*Bu), a kinetically stabilized member of the same class of compounds, could be achieved (Scheme 1) [3].

In general terms, tris(trimethylsilyl)phosphine (**1**) reacts with acyl chlorides **2** to furnish the phosphaalkenes **5**; the primarily formed carboxylic acid phosphides **3** can be detected by ^{31}P-NMR spectroscopy but undergo very rapid [1,3]-silyl shifts to generate the

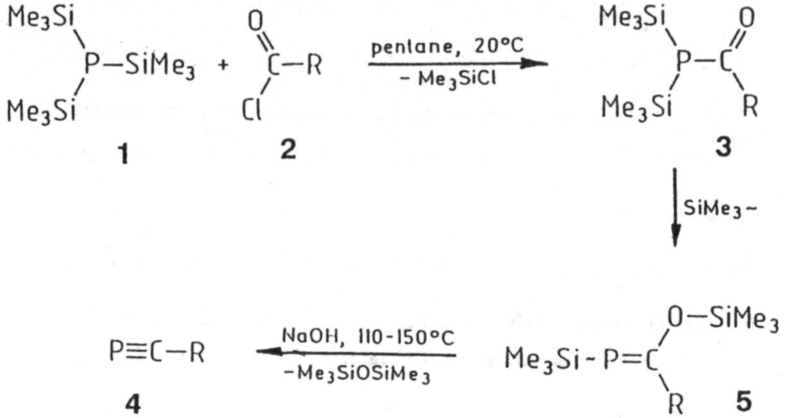

SCHEME 1

isomers **5** containing a P/C double bond system [4,5,6]. In contrast to Becker's original synthesis [3], our procedure for the elimination of hexamethyldisiloxane from **5** involves heating at 110-150 °C under vacuum over solid sodium hydroxide to furnish the target compound **4** which is subsequently separated from the hexamethyldisiloxane by distillation or fractional condensation [6,7,8].

2 Metal Complexes of Phosphaalkynes

The phosphaalkynes **4** have a very pronounced tendency to undergo cycloaddition reactions and thus resemble the alkynes [1,9, 10,11,12,13,14]. Thus, it is not surprising that they undergo cyclooligomerizations in the presence of organometallic auxiliary reagents in which the metallic fragment of the latter is incorporated in the product (Scheme 2) [9,15,16].

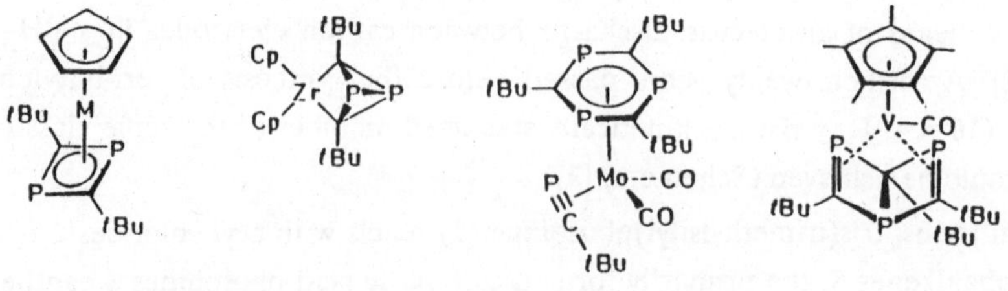

SCHEME 2

The rhodium [17] and cobalt [18] diphosphete derivatives **6** (M = Rh or Co) were the first dimeric complexes of **10** (≡ 4, R = *t*Bu) to be reported and contain two phosphaalkyne units. The inclusion of a Cp_2Zr fragment led to the completely new, trimeric phosphaalkyne complex **7** with a tricyclic structure [19]. Cyclotrimerization reactions provide an access to the complexed 1,3,5-triphosphabenzene system **8** [20] and to the Dewar benzene derivative **9** which is stabilized by the Cp*VCO fragment [21]. Furthermore, cyclooligomerization reactions of **10** (≡ 4, R = *t*Bu) with 1-methylnaphthalene(toluene)iron are now known and furnish sandwich complexes of four- and five-membered ring, phosphorus-containing ligands about a central iron atom [22].

3 Thermal Cyclotetramerization Reactions of 10 (≡ 4, R = *t*Bu)

Although the zirconium complex **7** plays a key role in the construction of the tetraphosphacubane system **11**, the purely thermal cyclooligomerization of the phosphaalkyne **10** (≡ 4, R = *t*Bu) will be discussed first (Scheme 3).

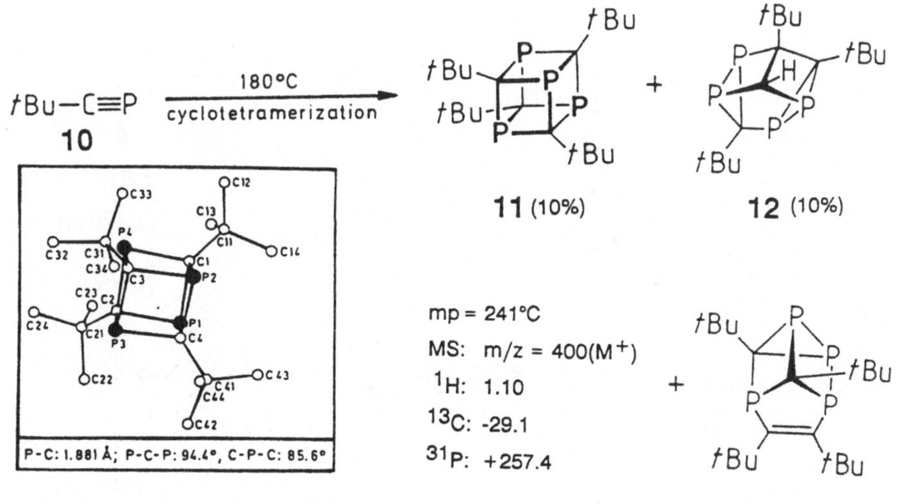

*t*Bu—C≡P **10** —180°C, cyclotetramerization→ **11** (10%) + **12** (10%) + **13** (10%)

P-C: 1.881 Å; P-C-P: 94.4°, C-P-C: 85.6°

mp = 241°C
MS: m/z = 400(M^+)
^1H: 1.10
^{13}C: -29.1
^{31}P: +257.4

SCHEME 3

The reaction proceeds at 130-180 °C but is not selective and furnishes, in addition to the tetraphosphacubane **11** [23], the tetraphosphacuneane **12** (one tert-butyl group is lost as iso-butylene) [24] and the tetraphosphabis(homo)prismane **13** [23, 24]. However, our interest will be directed solely to the highly symmetrical cubic system which exhibits unusual bonding characteristics that were previously unknown in phosphorus chemistry.

Thus, neither the ^{13}C-NMR signal of the carbon atoms in the cage of **11** at δ = -29.1 nor the ^{31}P-NMR signal shifted to very low field (d = +257.4) [23] can be explained by classical bonding concepts. However, PE spectroscopic investigations and MO model calculations have revealed that the "lone electron pairs" at phosphorus participate in the σ(P-C) bonds of the skeleton; this may be assumed to cause to a positive partial charge at phosphorus and a negative partial charge at carbon [25]. In this light, the extreme NMR chemical shifts are no longer surprising. Finally, an X-ray crystal structure analysis demonstrated a distorted cubic structure with equal P/C bond lengths (1.881 Å). As could be expected, the internal angles at carbon are somewhat increased (94.4°) while those at phosphorus are somewhat smaller (85.6°) [23].

4 Specific Synthesis of Tetraphosphacubane

Starting material for a chemoselective synthesis of the tetraphosphacubane **11** is the dimeric phosphaalkyne complex **15** (Scheme 4) [19].

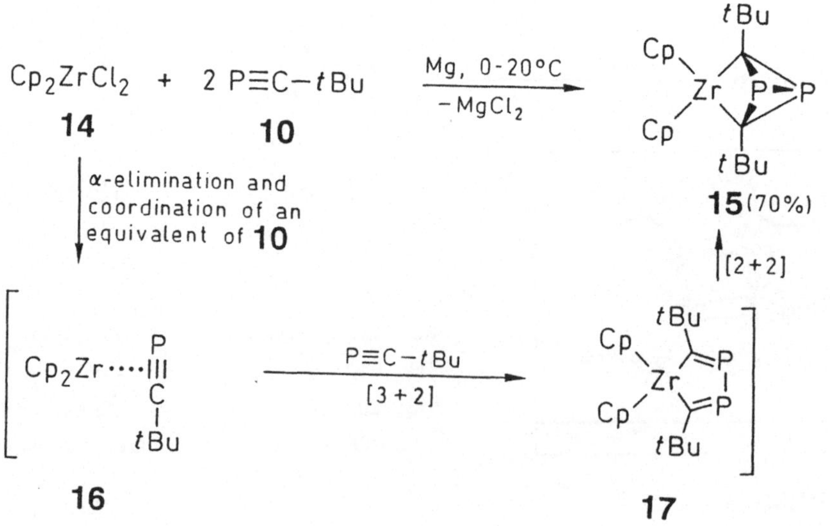

Cp_2ZrCl_2 + 2 P≡C−*t*Bu (**14**, **10**) —Mg, 0-20°C / $-MgCl_2$→ **15** (70%)

α-elimination and coordination of an equivalent of **10** → **16** $[Cp_2Zr \cdots P{\equiv}C{-}tBu]$ —P≡C−*t*Bu [3+2]→ **17** —[2+2]→ **15**

SCHEME 4

Compound **15** was obtained by treatment of bis(cyclopentadienyl)zirconium dichloride (**14**) with metallic magnesium in the presence of the phosphaalkyne **10** (≡ **4**, R = *t*Bu). It may be assumed that the metal is inserted into the Zr/Cl bond. which leads to elimination of magnesium chloride and generation of the coordinatively unsaturated species [Cp_2Zr]. "Side-on" coordination with one equivalent of **10** (≡ **4**, R = *t*Bu) gives rise to **16** which takes up a further equivalent of the phosphaalkyne to furnish **17**. The reaction sequence is terminated by an intramolecular, crossed [2+2]-cycloaddition yielding the tricyclic complex **15** [19]. The high field signal in the ^{31}P-NMR spectrum (δ = -247.0) is typical for the diphosphirane unit; last doubts on the structure of the complex were resolved by an X-ray crystal structure analysis [19].

The prerequisite for the conversion **15** ⟶ **11** is the removal of the Cp_2Zr fragment; the reactions illustrated in Scheme 5 show that this is, in principle, possible [26].

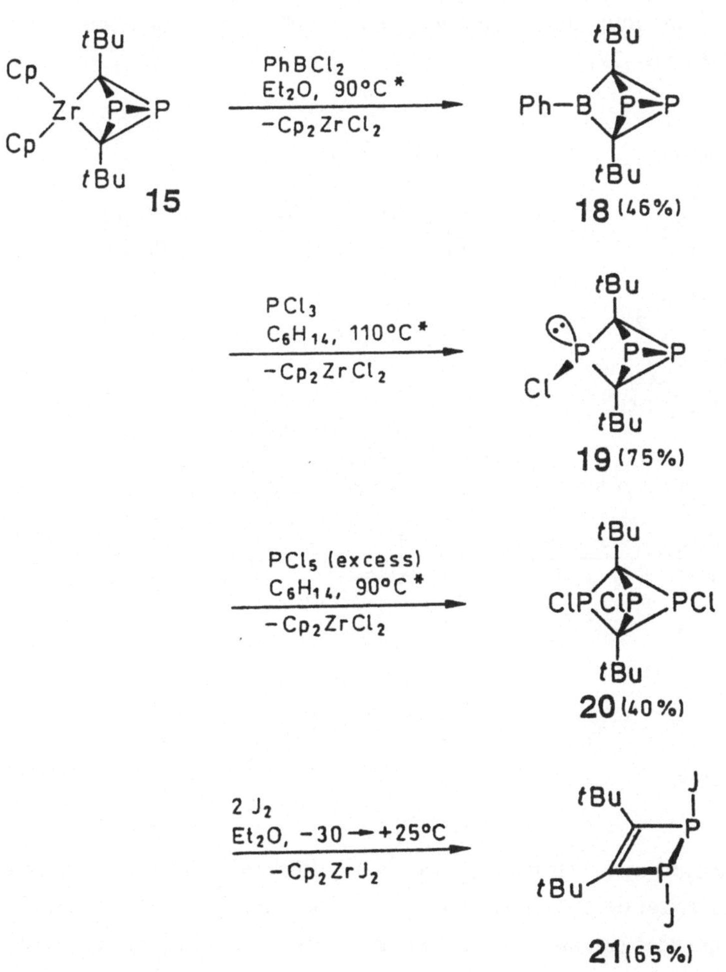

15 → **18** (46%): $PhBCl_2$, Et_2O, 90°C *, $-Cp_2ZrCl_2$

15 → **19** (75%): PCl_3, C_6H_{14}, 110°C *, $-Cp_2ZrCl_2$

15 → **20** (40%): PCl_5 (excess), C_6H_{14}, 90°C *, $-Cp_2ZrCl_2$

15 → **21** (65%): 2 J_2, Et_2O, −30 → +25°C, $-Cp_2ZrJ_2$

*) Schlenk pressure tube

SCHEME 5

The zirconium fragment can be exchanged by treatment with dichloro(phenyl)borane and trichlorophosphine, albeit under relatively harsh conditions, with retention of the tricyclic skeleton (**15** → **18** and **15** → **19**). In terms of energy, these reactions benefit from the Zr/Cl bond energies. The same type of reaction can also be achieved with phosphorus pentachloride but the excess of reagent employed results in cleavage of the P/P bond and simultaneous formation of two P/Cl bonds (**15** → **20**). Iodine is also able to remove the Cp_2Zr fragment but is concomitantly incorporated into the product (**15** → **21**); the trans-stereochemistry of the P/P bond was confirmed by an X-ray crystal structure analysis [26].

The objective of removing the Cp_2Zr fragment from **15** without inclusion of the halogenating reagent into the product was finally achieved by reaction with hexachloroethane (Scheme 6) [27].

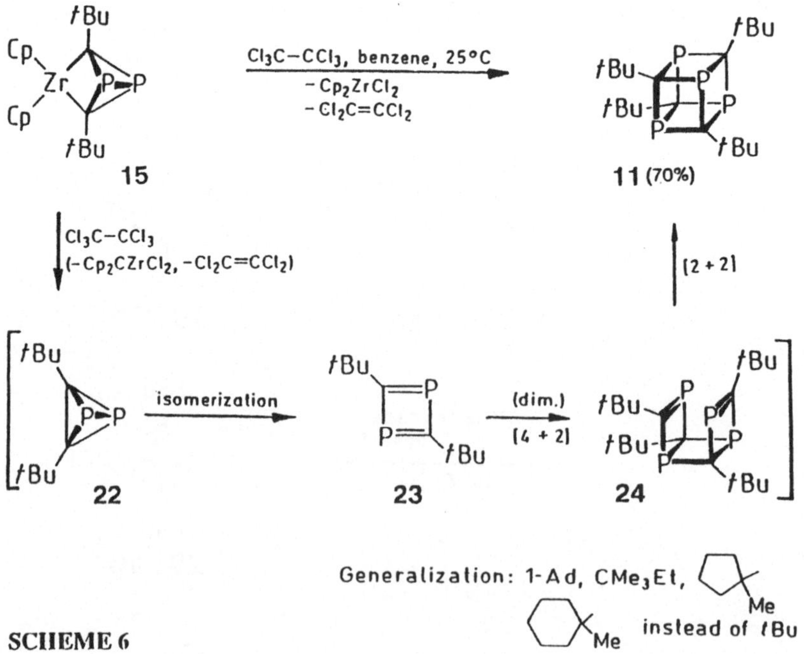

$Cl_3C{-}CCl_3$, benzene, 25 °C; $-Cp_2ZrCl_2$; $-Cl_2C{=}CCl_2$

15 → **11** (70%)

$Cl_3C{-}CCl_3$ ($-Cp_2CZrCl_2$, $-Cl_2C{=}CCl_2$)

22 —isomerization→ **23** —(dim.) [4 + 2]→ **24** —[2 + 2]→ **11**

Generalization: 1-Ad, CMe_3Et, 1-methylcyclopentyl, 1-methylcyclohexyl instead of *t*Bu

SCHEME 6

Irrespective of its actual structure, the primarily formed intermediate undergoes chemoselective dimerization to furnish tetraphosphacubane **11** in 70% yield [27]. Analogous reaction sequences can also be realized via the corresponding tricyclic zirconium complexes with the phosphaalkynes **4** (R = 1-adamantyl, tert-pentyl, 1-methyl-1-cyclopentyl, and 1-methyl-1-cyclohexyl) [27,28].

A feasible explanation for the outcome of the reaction involves the intermediate formation of the diphosphatetrahedrane **22** which isomerizes to the 1,3-diphosphete **23**. Dimerization of the latter in the sense of a hetero-Diels-Alder reaction furnishes the tetraphosphatricyclooctadiene **24** which, in turn, undergoes an intramolecular [2+2]-cycloaddition to provide the product **11** [27].

The fact that the zirconium complex **15** can indeed be converted into the diphosphete **23** with elimination of the Cp_2Zr fragment upon reaction with hexachloroethane is demonstrated by the following experiment (Scheme 7) [27].

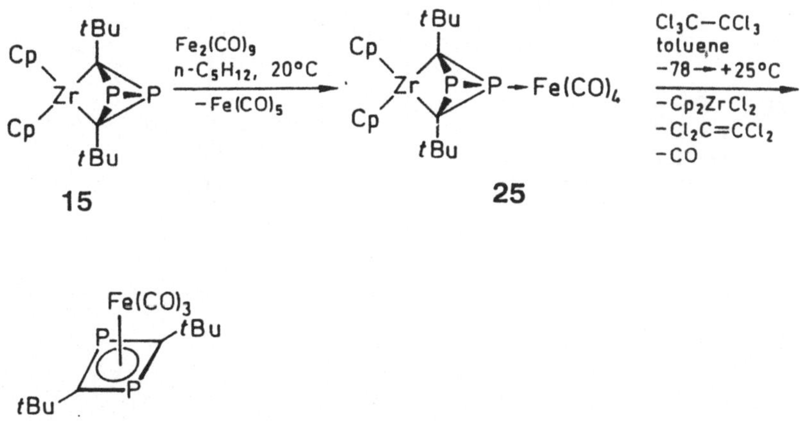

tBu, Cp, Cp, Zr, P, P, tBu — **15** — $Fe_2(CO)_9$, n-C_5H_{12}, 20°C, $-Fe(CO)_5$ → tBu, Cp, Cp, Zr, P, P–$Fe(CO)_4$, tBu — **25** — Cl_3C-CCl_3, toluene, −78 → +25°C, $-Cp_2ZrCl_2$, $-Cl_2C{=}CCl_2$, −CO →

$Fe(CO)_3$, P, tBu, P, tBu — **26**

SCHEME 7

When **15** is allowed to from a complex with nonacarbonyldiiron, the tetracarbonyliron complex **25** is obtained (^{31}P-NMR: δ = -134.4, -272.5; $^1J_{P,P}$ = 106.8 Hz). Treatment of this complex with the chlorinating agent does indeed result in the formation of the complexstabilized diphosphete **26** [27]. This product is identical with that obtained from the reaction of the phosphaalkyne **10** (≡ 4; R = *t*Bu) with $Fe_2(CO)_9$ [29].

5 Functionalization of the Tetraphosphacubane 11

No reactions of the other tetraphosphacubanes **11** (SnPh [30], Si-*t*Bu [31], or Al [32] in place of C-*t*Bu) have, as yet been reported. However, the simple access **15** ⟶ **11** provides the first possibility to investigate functionalization reactions at phosphorus. Since the heteroatoms in **11** - as reported [25] - have decreased nucleophilicity as a result of the participation of the lone pairs of electrons in the P/C σ-bonds, some of the reactions with electrophiles require rather drastic reaction conditions [33].

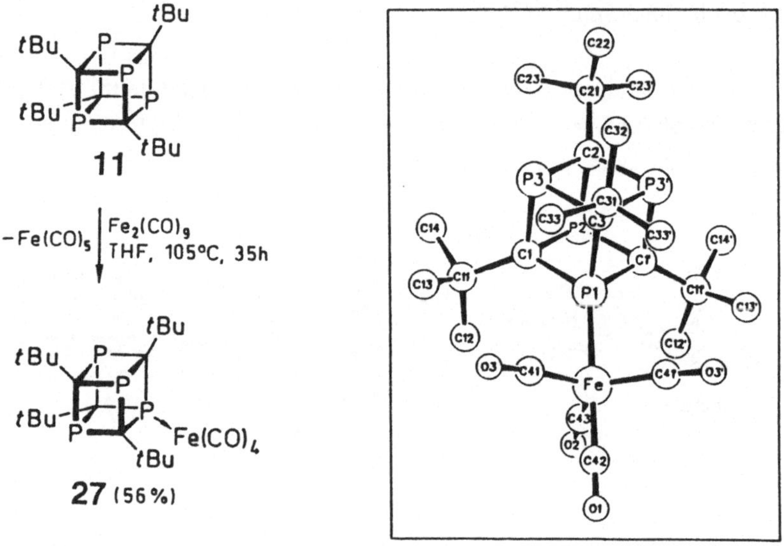

SCHEME 8

When **11** is heated with $Fe_2(CO)_9$ in tetrahydrofuran, the first metal complex (**27**) of a tetraphosphacubane is obtained. Even when two molar equivalents of the iron carbonyl are employed, only one $[Fe(CO)_4]$ fragment is attached to **11**. The NMR signals of one skeletal carbon atom and the complexed phosphorus atom remain unchanged (relative to **11**) upon complex formation. The structure of **27** (m symmetry) was unequivocally substantiated by an X-ray crystal structure analysis [33]. Worthy of note are the enlargements of the exocyclic angles P1-C1-C11 (≡ P1-C1'-C11') and P1-C3-C31 to 131.1(3)° and 132.5(4)° in comparison to **11** (122 °C [23]). This may be attributed to steric interactions between the tert-butyl groups and the $[Fe(CO)_4]$ fragment. This effect, which will be mentioned again below, presumably prevents the addition of a further metal fragment [33].

Methyl iodide is not able to quaternize **11**. However, stronger alkylating agents such as methyl trifluoromethanesulfonate and ethynyl(phenyl)iodonium triflate [34] do react with this pentacyclic compound but, again, only monoalkylation is observed even with excess reagent (**11** → **28** and **11** → **29**; Scheme 9) [33,35].

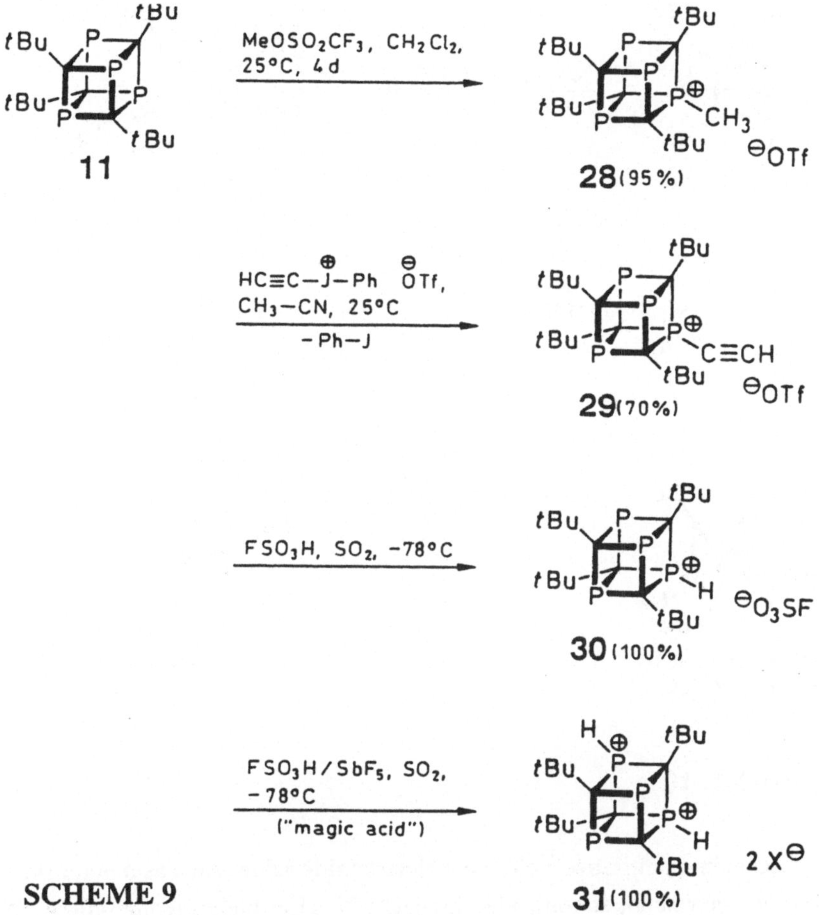

$MeOSO_2CF_3$, CH_2Cl_2, 25°C, 4d → **28** (95%)

$HC{\equiv}C{-}\overset{\oplus}{J}{-}Ph$ $\overset{\ominus}{O}Tf$, $CH_3{-}CN$, 25°C, − Ph−J → **29** (70%)

FSO_3H, SO_2, −78°C → **30** (100%)

FSO_3H/SbF_5, SO_2, −78°C ("magic acid") → **31** (100%)

SCHEME 9

The same is valid for the protonation of **11** by fluorosulfonic acid in liquid sulfur dioxide at low temperature (**11** → **30**) [35]. Only "magic acid" is able to effect double protonation of the pentacyclic compound (**11** → **31**). The above two observations have been confirmed unambiguously by NMR investigations [35]. The results of hydrolysis experiments which regenerate the tetraphosphacubane **11** from the mono- and diprotonated species are in complete accord with the above observations [35].

The Staudinger reaction of **11** with azides, however, results in the formation of iminophosphoranes with 1:1 and 1:2 stoichiometries (Scheme 10) [33,36].

p-Tosyl azide, a highly electrophilic reagent, undergoes addition to **11** already at room temperature to furnish the stable triazene **32**. When this product is stored in chloroform (which, of course, contains protons), the expected extrusion of nitrogen

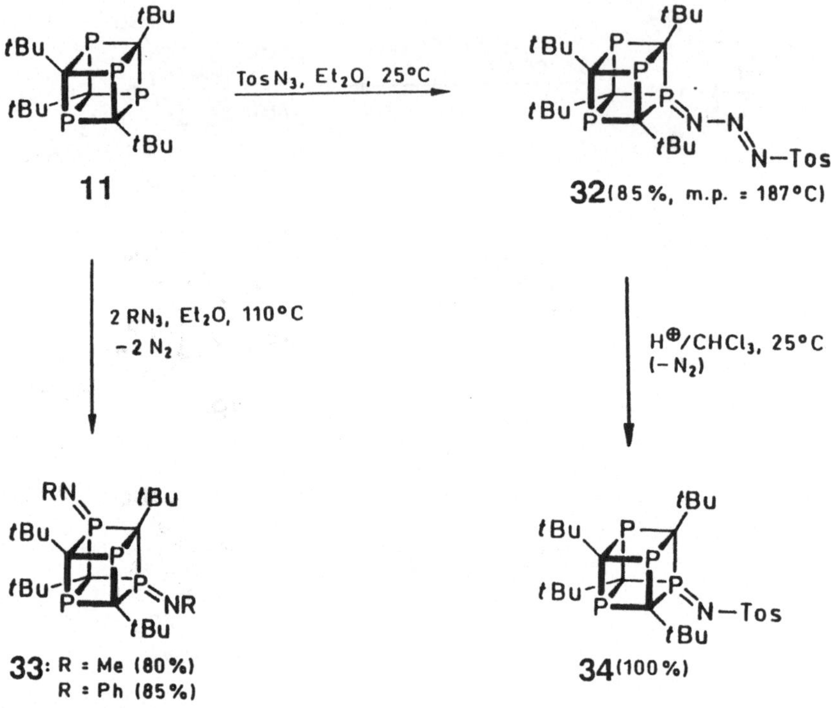

11 → ($TosN_3$, Et_2O, 25°C) → **32** (85 %, m.p. = 187°C)

11 → (2 RN_3, Et_2O, 110°C, $-2\ N_2$) → **33**: R = Me (80 %), R = Ph (85 %)

32 → ($H^{\oplus}/CHCl_3$, 25°C, $(-N_2)$) → **34** (100 %)

SCHEME 10

(⟶ **34**) occurs in quantitative yield. Less electrophilic azides such as methyl and phenyl azide, on the other hand, are only able to react with **11** at higher temperatures (Schlenk pressure tube technique). In the presence of at least two equivalents of the azide, the bis(iminophosphoranes) **33** are formed selectively [33].

Reactions of diazo compounds, which are isoelectronic with azides, with the tetraphosphacubane **11** have, as yet, only resulted in the formation of phosphazines (Scheme 11). The formation of a methylenephosphorane via nitrogen elimination has not yet been reported [33].

Diazomethyl compounds bearing a iPr, *t*Bu, Ph, or CO_2Et substituent and also those with two sterically demanding substituents such as diazo(diisopropyl)methane and diazo(diphenyl)methane react with **11** in a molar ratio of 1:1 to furnish the monophosphazines **35** in high yields (68-90%) [36]. Under comparable reaction conditions, the incorporation of a second phosphazine group (⟶ **36**) is only possible with monosubstituted diazoalkanes (R^1 = H, R^2 = *i*Pr, *t*Bu, Ph; 85% yields) [33,36].

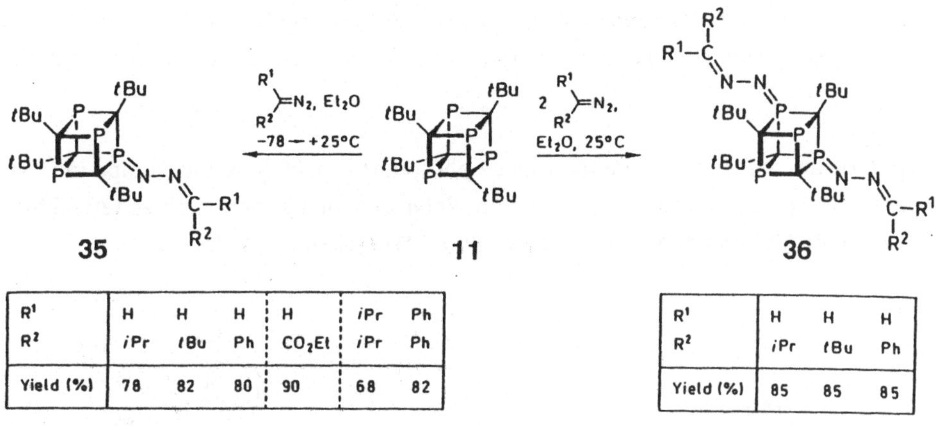

R^1	H	H	H	H	*i*Pr	Ph
R^2	*i*Pr	*t*Bu	Ph	CO_2Et	*i*Pr	Ph
Yield (%)	78	82	80	90	68	82

R^1	H	H	H
R^2	*i*Pr	*t*Bu	Ph
Yield (%)	85	85	85

SCHEME 11

The coupling of two phosphacubane units through a bridging group can be achieved by both the Staudinger reaction with bifunctional azides and the phosphazine formation process with bifunctional diazo compounds (Scheme 12) [33].

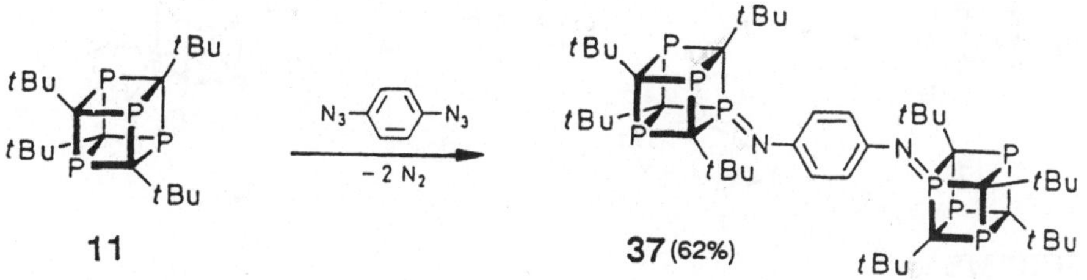

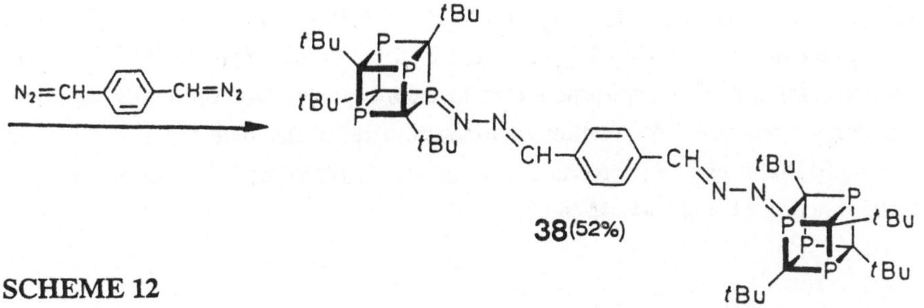

SCHEME 12

Thus, **11** reacts with 1,4-diazidobenzene - albeit at a reaction temperature of 110 °C - in a 1:2 process to furnish the bis-(iminophosphorane) **37**. In contrast to this reaction, the "coupling" of the same tetraphosphacubane with 1,4-bis-(diazomethyl)benzene occurs at room temperature to produce the product **38** by way of double phosphazine formation [33].

In spite of the reduced electron density at phosphorus in **11** [25], this compound does undergo [4+1]-cycloaddition reactions with electron-poor partners such as tetrachloro-orthoquinone (**39**) and diethyl azodicarboxylate (**41**) (Scheme 13) [33].

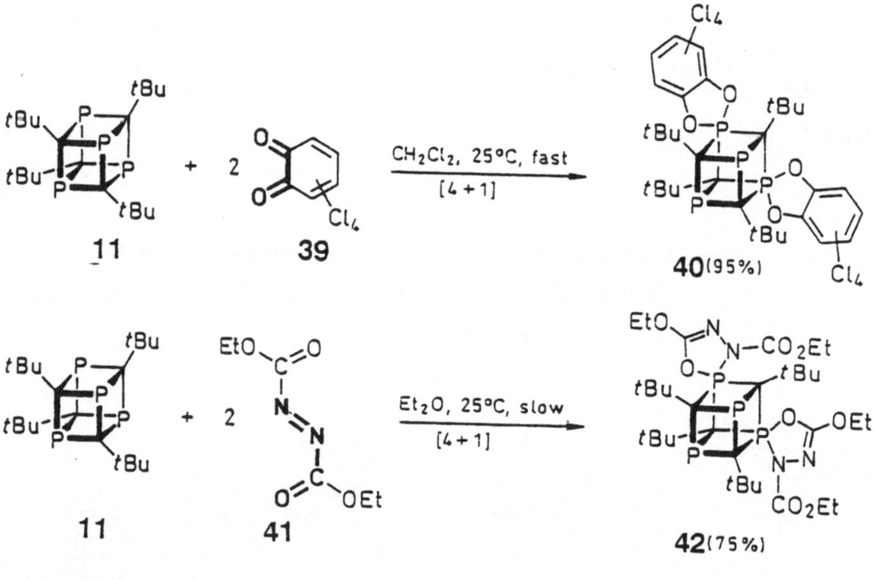

SCHEME 13

Both cycloaddition processes proceed at room temperature and give rise to the bis(spiro) compounds **40** and **42**, respectively. Even with a large excess of the hetero-1,3-diene, the reaction does not go beyond the 1:2 stoichiometry. It is readily apparent that the two remaining $\lambda^3\sigma^3$-phosphorus atoms in each cage are now optimally shielded and thus no longer accessible for further cycloaddition reactions. The reaction **11** + **39** ⟶ **40** is so rapid that it can be performed as a sort of "titration" and thus differs significantly from the reaction **11** + **41** ⟶ **42** [33,36].

With sterically less demanding reaction partners such as sulfur (in the presence of triethylamine), the pentacyclic system **11** can be oxidized stepwise at phosphorus until all four phosphorus atoms have reacted (Scheme 14) [36].

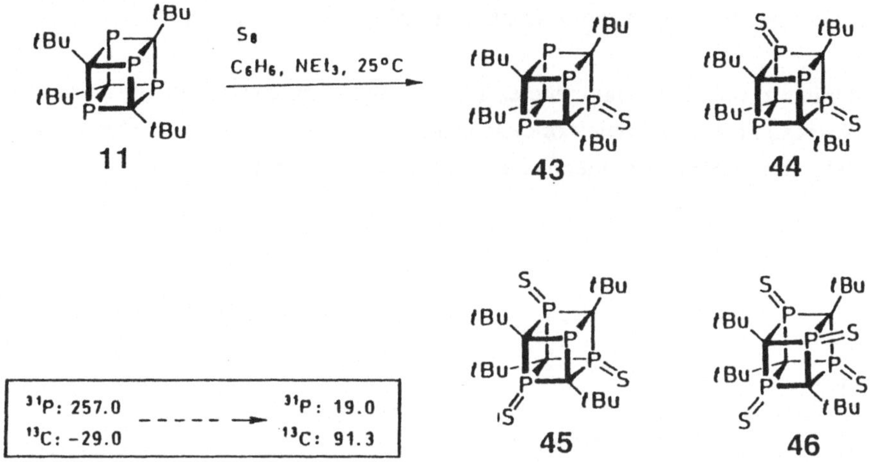

SCHEME 14

In the final analysis, the product spectrum merely reflects the ratio of the two reaction partners. Thus, the mono-, bis-, tris-, and tetra(thioxo)tetraphosphacubanes **43-46** can each be obtained in the pure state (in some cases after chromatographic work-up) [36]. In the case of the reaction **11** → **46** as a typical example, the dramatic changes in the bonding situation among the cage atoms can be demonstrated clearly by ^{13}C- and ^{31}P-NMR spectroscopy: the low field shift of the carbon signal from δ = -29.0 to δ = +91.3 not only shows the normality of the situation but is also indicative of the acceptor character of the P=S groups. At the same time, the phosphorus signal experiences a dramatic shift to high field (up to δ = +19.0, an acceptable value for $\lambda^5\sigma^4$-phosphorus) [36].

When **11** is treated with four equivalents of bis(trimethylsilyl) peroxide in dichloromethane at room temperature, the stable tetraoxide **46** (O in place of S) is obtained. The NMR signals for the cage atoms show similar trends as those of the tetrasulfide [33]. It is interesting to note that, even with a large excess of elemental selenium, the tetraselenide **46** (Se in place of S) cannot be obtained; the reaction stops at the stage of the triselenide **45** (Se in place of S). A crystal structure analysis of this product revealed that the exocyclic P-C-C bond angles at the unaffected phosphorus atom are 118°, i.e. they are reduced by 4° in comparison to those of **11** (122°). This apparently prevents the introduction of the fourth selenoxo group for steric reasons [33].

6 Spirocyclooligomerization of Phosphaalkynes with AlX_3

In addition to the zirconium complex **15**, the aluminium compound **50** is also of major significance in the construction of cyclooligomers of phosphaalkynes (Scheme 15), since, as will be shown below, the metal can be removed easily [37].

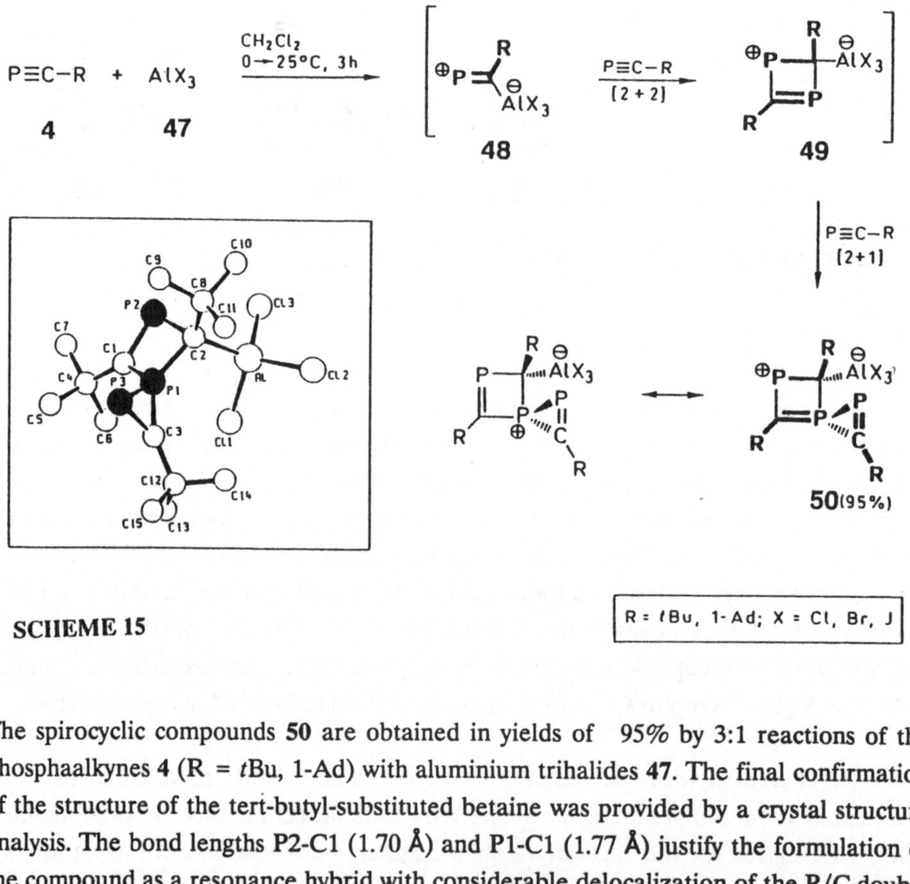

P≡C−R + AlX_3 (4, 47) —CH_2Cl_2, 0→25°C, 3h→ [48] —P≡C−R [2+2]→ [49] —P≡C−R [2+1]→ 50 (95%)

R = *t*Bu, 1-Ad; X = Cl, Br, J

SCHEME 15

The spirocyclic compounds **50** are obtained in yields of 95% by 3:1 reactions of the phosphaalkynes **4** (R = *t*Bu, 1-Ad) with aluminium trihalides **47**. The final confirmation of the structure of the tert-butyl-substituted betaine was provided by a crystal structure analysis. The bond lengths P2-C1 (1.70 Å) and P1-C1 (1.77 Å) justify the formulation of the compound as a resonance hybrid with considerable delocalization of the P/C double bond or, respectively, a positive charge on the hetero-atom according to **50**. The two values for the bond lengths are larger than the average value for open-chain phosphaalkenes [38] but are still much less than the values for P/C single bonds [37].

We assume that the Lewis acids **47** add to the carbon atom of the phosphaalkynes in the primary step (→ **48**) [39], which corresponds to the polarization of the triple bond [10]. This Lewis acid adduct then undergoes a [2+2]-cycloaddition with a second equivalent of **4** to furnish the dimer complex **49** [40]. A final [2+1]-cycloaddition of the latter with a third equivalent of phosphaalkyne is then responsible for the end product formed (**49** → **50**) [37].

When the spirocyclic betaine **50** (R = *t*Bu, X = Cl) is treated with the Lewis base DMSO, the diphosphete **51** is formed under concomitant loss of aluminium trichloride. This species contains both $\lambda^{3}{}_{\sigma}{}^{2}$- and $\lambda^{5}{}_{\sigma}{}^{4}$-phosphorus atoms. It is unstable even at -45 °C and rearranges with cleavage of the P/P bond in the diphosphirene ring to furnish the Dewar 1,3,5-triphosphabenzene **52**. However, since the bicyclic compound **52** also cannot be isolated, a fourth equivalent of **4** (R = *t*Bu; ≡ **10**) is added to the reaction mixture before generation of **51**. This then participates in a homo-Diels-Alder reaction with **52** to furnish the phosphaalkyne cyclotetramer **54** [37].

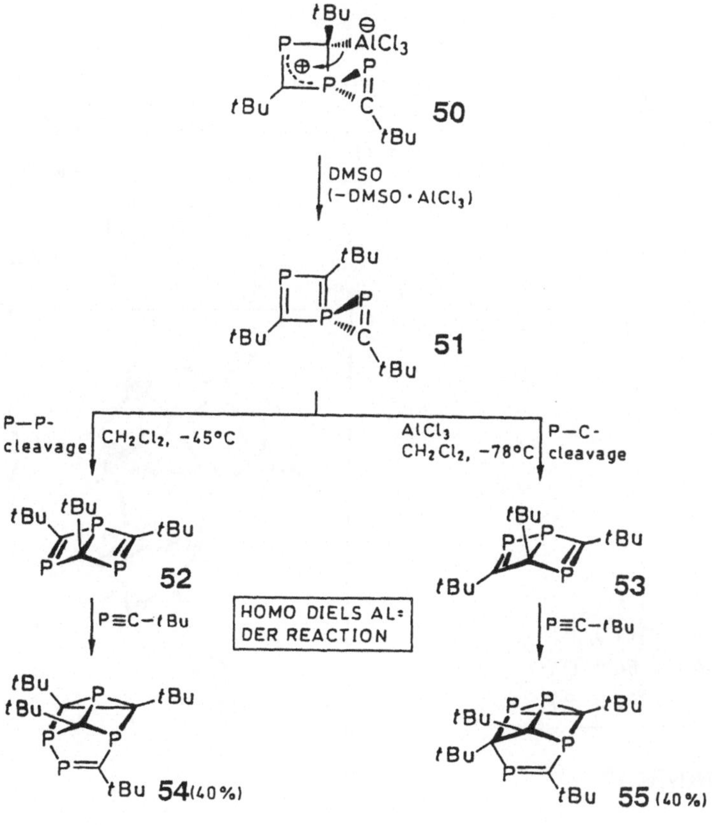

SCHEME 16

When the same experiment is carried out with an excess of aluminium trichloride, **51** rearranges with cleavage of the P/C bond in the diphosphirene ring to generate the Dewar 1,2,5-triphosphabenzene **53** which can also be trapped by a homo-Diels-Alder reaction (→ **55**) [37]. The role of the Lewis acid (or the $AlCl_3$·DMSO adduct, respectively) in the isomerization reaction is still not clear. Both processes are highly selective and allow the specific construction of the isomeric, polycyclic products **54** and **55**.

The structural elucidation of the phosphaalkyne cyclotetramers **54** and **55** was first based on the chemical shifts of the skeletal atoms and the coupling constants caused by phosphorus [37]; absolute certainty, at least with regard to the tetraphosphatetracyclic system **55**, was established from the X-ray crystal structure analysis of the cycloadduct **56**, prepared by reaction of **55** with mesitylnitrile oxide (Scheme 56).

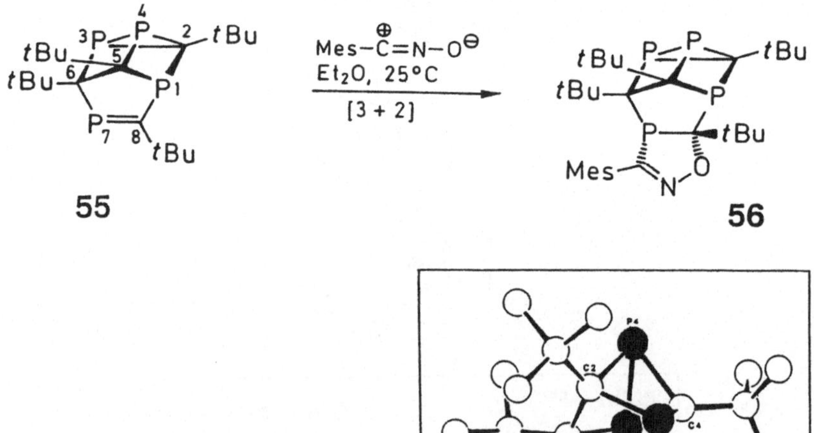

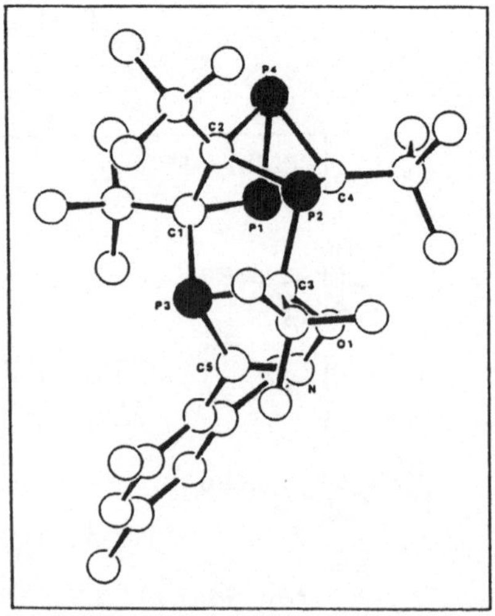

^{31}P: δ = −174.4, P4
δ = −147.3, P3
δ = + 64.6, P1
δ = +399.0, P7

SCHEME 17

The addition of the dipole to the P/C bond of **55** proceeded regio- and stereospecifically whereas, on the other hand, a comparable reaction with **54** could not be realized. The markedly different P/C bond lengths, varying between 1.807 (P2-C4) and 1.927 Å (P4-C2), are worthy of note in the structure of **56** [37].

7 Isomerization Reactions among the Compounds 13, 54, 55, and 57

The phosphaalkyne cyclotetramers **13**, **54**, **55**, and **57** undergo interesting, mutual isomerization reactions under thermal or photochemical conditions; the tetraphosphacubane **11**, however, does not participate in these reactions (Scheme 18) [24,41].

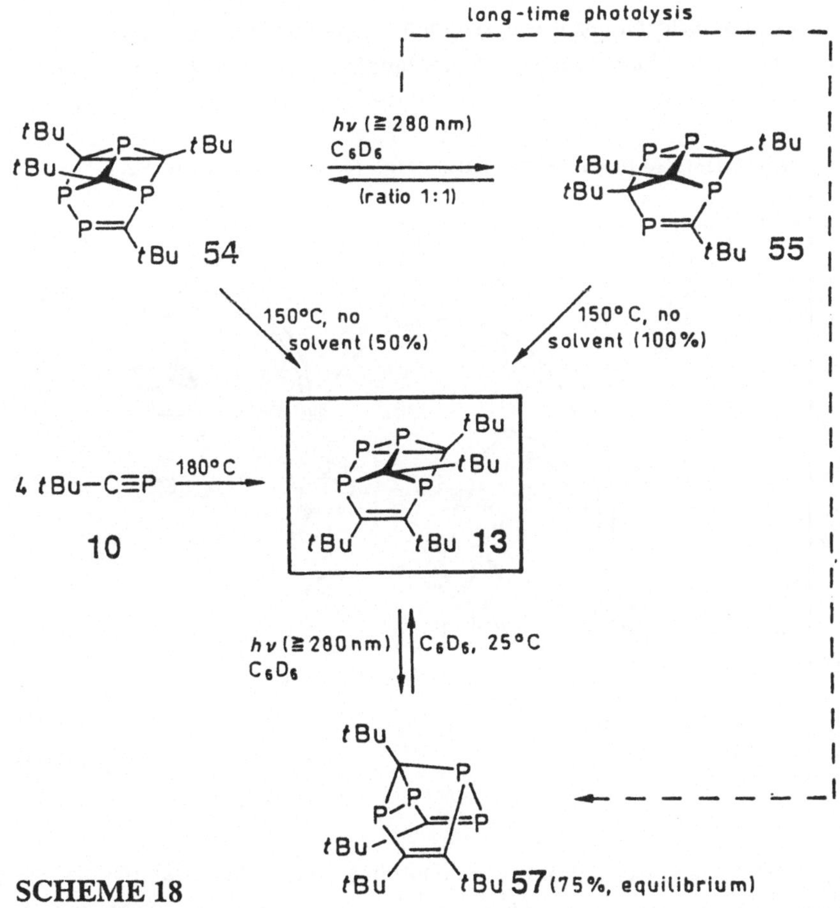

SCHEME 18

Thus, on heating to 150 °C, both **54** and **55** are converted to the isomer **13** [41]; the latter can also be obtained directly by pyrolysis of the phosphaalkyne **10** (≡ **4**; R = *t*Bu) [24].

In addition, both pure isomers **54** and **55** can be induced to form a 1:1 equilibrium mixture by application of photochemical conditions [41]. On long duration photolysis of the 1:1 equilibrium mixture of **54** and **55**, the tetraphosphasemibullvalene **57** is formed as a new isomer. This compound is also formed in limited amounts during the photolysis of **13** and, under thermal conditions (25 °C), **57** is converted back to **13** [24]. No investigations on the occurrence of intermediates in these isomerization processes have yet been reported.

The calculated heats of formation (ΔH_f) for the unsubstituted parent compounds are in accord with the above-mentioned, mutual transformations in the system **13/54/55/57**. The calculations were performed using the semi-empirical PM3 programme (Scheme 19) [41]. In these calculations it was assumed throughout that the replacement of a hydrogen atom by a tert-butyl group would not fundamentally change the situation.

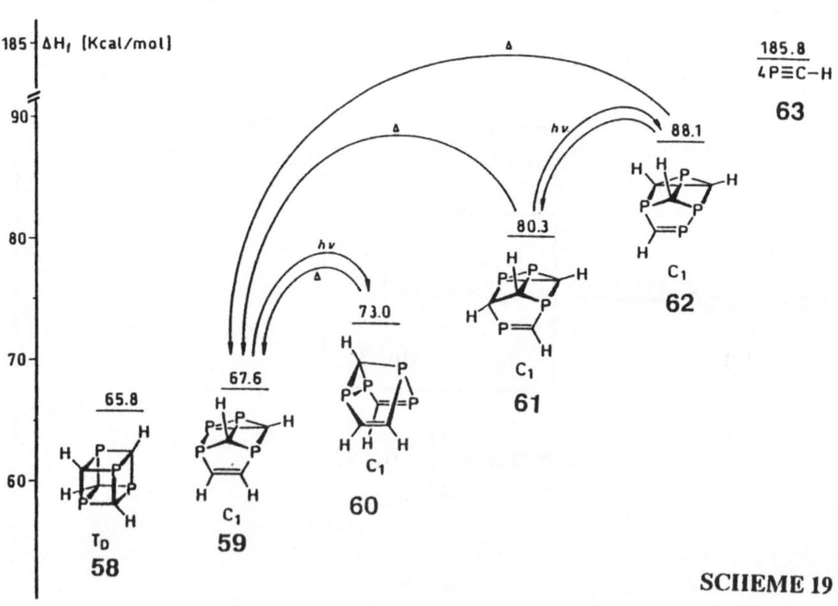

SCHEME 19

Starting point of these calculations is the phosphaacetylene **63** [2,42]; the ΔHf value for four such molecules amounts to 185.8 kcal/mol. Of course, the cyclotetramerization leads in every case to a drastic decrease in the energy content. The unsubstituted tetraphosphacubane **58** lies at the other end of the scale, preceded by the isomer **59**. It is clear that, in the case of tert-butyl substitution, the latter species is accessible from **60**,

61, and **62** under thermal conditions. The conversion **62** → **61** is equally reasonable. On the other hand, the photochemical transformations **59** → **60** and **61** → **62** in which the products and substrates lie relatively close together with regard to their DHf values are also not unexpected [41].

Acknowledgements

I am grateful to the Deutsche Forschungsgemeinschaft, the Volkswagen-Stiftung, the Fonds der Chemischen Industrie, and the Landesregierung von Rheinland-Pfalz for generous financial support of our own investigations mentioned in this review.

References

[1] Organophosphorus Compounds; Part 64. For Part 63, see: H. Memmesheimer, M. Regitz in U. Brinker (Ed.): Advances in Carbene Chemistry, JAI Press, Greenwich, Connecticut 1992, in press.

[2] T. E. Gier, J. Am. Chem. Soc. 63(1961)1769.

[3] G. Becker, G. Gresser, W. Uhl, Z. Naturforsch. 36b(1981)16.

[4] G. Becker, Z. Anorg. Allg. Chem. 430(1977)66.

[5] T. Allspach, M. Regitz, G. Becker, W. Becker, Synthesis 1986, 31.

[6] W. Rösch, U. Vogelbacher, T. Allspach, M. Regitz, J. Organomet. Chem. 306(1986)39.

[7] M. Regitz, W. Rösch, T. Allspach, U. Annen, K. Blatter, J. Fink, M. Hermesdorf, H. Heydt, U. Vogelbacher, O. Wagner, Phosphorus Sulfur 30(1987)479.

[8] W. Rösch, U. Hees, M. Regitz, Chem. Ber. 120(1987)1645.

[9] Review: M. Regitz, P. Binger, Angew. Chem. 100(1988)1541; Angew. Chem., Int. Ed. Engl. 27(1988)1484.

[10] Review: M. Regitz, Chem. Rev. 90(1990)191.

[11] Report: M. Regitz, Nachr. Chem. Tech. Lab. 37(1989)896.

[12] Report: M. Regitz in E. Block (Ed.): Heteroatom Chemistry, VCH Publisher, New York 1990, p. 295 ff.

[13] Report: M. Regitz, Bull. Soc. Chim. Belg. 101(1992)359.

[14] Monograph: M. Regitz in M. Regitz, O. J. Scherer (Eds.): Multiple Bonds and Low Coordination in Phosphorus Chemistry, Thieme, Stuttgart 1990, p. 58 ff.

[15] Cf. P. Binger, in Ref. [14], p. 90 ff.

[16] Review: J. F. Nixon, Chem. Rev. 88(1988)1327.

[17] P. B. Hitchcock, U. J. Maah, J. F. Nixon, J. Chem. Soc., Chem. Commun. 1986, 737.

[18] P. Binger, R. Milczarek, R. Mynott, M. Regitz, W. Rösch, Angew. Chem. 98(1986)645; Angew. Chem., Int. Ed. Engl. 25(1986)644; P. Binger, R. Milczarek, R. Mynott, C. Krüger, Y.-H. Tsay, E. Raabe, M. Regitz, Chem. Ber. 121 (1988) 637.

[19] P. Binger, B. Biedenbach, C. Krüger, M. Regitz, Angew. Chem. 99(1987)798; Angew. Chem., Int. Ed. Engl. 26(1987)764.

[20] A. R. Barron, A. H. Cowley, Angew. Chem. 99(1987)956; Angew. Chem., Int. Ed. Engl. 26(1987)907. All attempts to repeat this experiment to date have been unsuccessful (P. Binger, personal communication, Max-Planck-Institut für Kohlenforschung, Mülheim/Ruhr).

[21] P. Milczarek, W. Rüsseler, P. Binger, K. Jonas, K. Angermund, C. Krüger, M. Regitz, Angew. Chem. 99(1987)957; Angew. Chem., Int. Ed. Engl. 26(1987)908.

[22] M. Driess, D. Hu, H. Pritzkow, H. Schäufele, U. Zenneck, M. Regitz, W. Rösch, J. Organomet. Chem. 334(1987)C35.

[23] T. Wettling, J. Schneider, O. Wagner, C. G. Kreiter, M. Regitz, Angew. Chem. 101(1989)1035; Angew. Chem., Int. Ed. Engl. 28(1989)1013.

[24] T. Wettling, M. Regitz, unpublished results, University of Kaiserslautern 1990.

[25] R. Gleiter, K.-H. Pfeifer, M. Baudler, G. Scholz, T. Wettling, M. Regitz, Chem. Ber. 123(1990)757.

[26] P. Binger, T. Wettling, R. Schneider, F. Zurmühlen, U. Bergsträßer, J. Hoffmann, G. Maas, M. Regitz, Angew. Chem. 103(1991)208; Angew. Chem., Int. Ed. Engl. 30(1991)207.

[27] T. Wettling, B. Geißler, R. Schneider, S. Barth, P. Binger, M. Regitz, Angew. Chem. 104(1992)761; Angew. Chem., Int. Ed. Engl. 31(1992)758.

[28] B. Geißler, M. Regitz, unpublished results, University of Kaiserslautern 1991.

[29] P. Binger, B. Biedenbach, R. Schneider, M. Regitz, Synthesis 1989, 960.

[30] H. Schumann, H. Benda, Angew. Chem. 80(1968)846; Angew. Chem., Int. Ed. Engl. 7(1968)813.

[31] M. Baudler, G. Scholz, K.-F. Tebbe, M. Feher, Angew. Chem. 101(1989)352; Angew. Chem. Int. Ed. Engl. 28(1989)339.

[32] A. H. Cowley, R. A. Jones, P. R. Harris, D. A. Attwood, L. Contreras, C. J. Burek, Angew. Chem. 103(1991)1164; Angew. Chem., Int. Ed. Engl. 30(1991)1143.

[33] M. Birkel, J. Schulz, U. Bergsträßer, M. Regitz, Angew. Chem. 104(1992)870; Angew. Chem., Int. Ed. Engl. 31(1992)879.

[34] Review: P. Stang, Angew. Chem. 104 (1992) 281; Angew. Chem., Int. Ed. Engl. 31(1992)274.

[35] K. K. Laali, M. Regitz, M. Birkel, P. Stang, C. Crittel, J. Org. Chem. 57(1993) in press.

[36] M. Birkel, M. Regitz, unpublished results, University of Kaiserslautern 1992.

[37] B. Breit, U. Bergsträßer, G. Maas, M. Regitz, Angew. Chem. 104(1992)1043; Angew. Chem., Int. Ed. Engl. 31(1992)1055.

[38] R. Appel, in Ref. [14], p. 160.

[39] Compound **4** (R = *t*Bu) was protonated at the same carbon atom by fluorosulfonic acid in liquid SO_2: K. Laali, M. Regitz, unpublished results, Kent State University 1992.

[40] P. B. J. Driesen, H. Hogeveen, J. Organomet. Chem. 156(1978)265.

[41] B. Breit, H. Heydt, M. Regitz, unpublished results, University of Kaiserslautern 1992.

[42] E. P. O. Fuchs, M. Hermesdorf, W. Schnurr, W. Rösch, H. Heydt, M. Regitz, J. Organomet. Chem. 338(1988)329.

Metal Containing Compounds: Precursors for new Reactions and Materials

Herbert W. Roesky

Institut für Anorganische Chemie der Universität Göttingen, Tammannstraße 4, D-W-3400 Göttingen

Summary

The syntheses of six-membered metallacyclophosphazenes and metal containing siloxanes are described. Phosphazene and siloxane groups are isoelectronic a concept aiding in the synthesis of the target molecules. The metal containing silicones function as model compounds for metal oxides on silica surfaces. A first example of a metal containing borazine is given. Some of the compounds function as precursors for chemical vapor deposition.

1 Introduction

The six-membered $[NP(R_2)]_3$ (**1**) and $[OSi(R_2)]_3$ (**2**) ring systems are isoelectronic. Both have an extensive chemistry forming differently substituted molecules as well as polymeric materials.

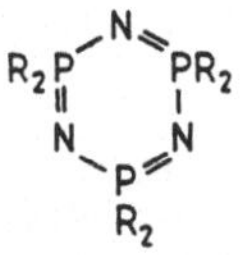

1 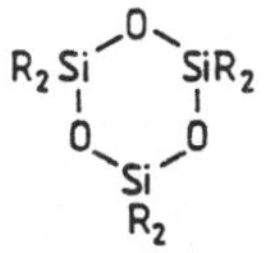 **2**

Innumerable derivatives of both the phosphazenes and the siloxanes have been prepared during the last 150 years [1]. In contrast, metal containing phosphazene and siloxane ring systems are rare. A number of metal containing four- to twelve-membered ring systems will be described.

Another well known ring system, the inorganic analog of benzene, $(HBNH)_3$ (**3**), a compound commonly called borazine, was prepared by Stock et al. [2] 76 years ago.

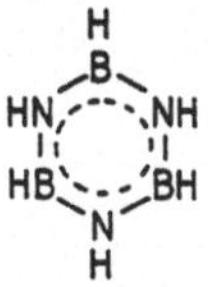

3

While a few metal containing benzene derivatives are known, metal containing borazines have been prepared only recently.

Particular emphasis will be given to organometallic oxides as model compounds for metal oxides on silica surfaces.

2 Metal Containing Cyclophosphazenes

In 1986 we reported on the preparation and structural investigation of the first six-membered metallacyclophosphazenes [3]. We have developed the following routes [1] for the preparation of molecules **4** - **7**.

1. $[H_2NPPh_2NPPh_2NH_2]Cl + WCl_6 \longrightarrow$ **4**

2. $[H_2NPPh_2NPPh_2NH_2]Cl + N{\equiv}MoCl_3 \longrightarrow$ **5**

3. $(CF_3)_2P(Cl){=}NSiMe_3 + Me_3SiN{=}VCl_3 \longrightarrow$ **6**

4. $Ph_2P(Cl) = NSiMe_3 + V(O)Cl_3 \longrightarrow$ **7**

All these ring systems **4** - **7** might be considered to be fused from phosphazenes and metal halides in high oxidation states. The common features of these systems are the metal halide bonds. Furthermore it turned out that nucleophilic substitution reactions often resulted in unpredicted products.

The six-membered rings can be opened to generate polymers. However, the polymers contain hydrolytically unstable metal halide bonds. Therefore, we were interested in synthesizing systems containing phosphazene and metal oxides. A straightforeward reaction was developed to the following equation.

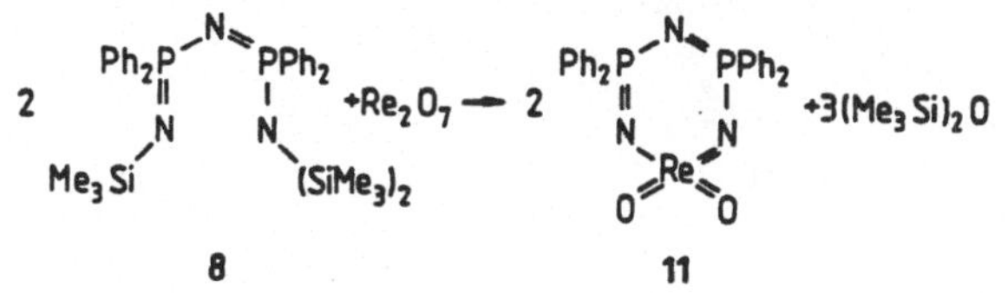

Compound **8** was prepared from $[H_2NPPh_2NPPh_2NH_2]Cl$ and KH or NEt_3 to yield $H[N(PPh_2NH)_2]$ the dehydrohalogenated intermediate. Treatment of the intermediate with Me_2NSiMe_3 resulted in the formation of $H_2NPPh_2NPPh_2NSiMe_3$ **9** or $HN(PPh_2NSiMe_3)_2$ **10**. This is only dependant on the ratio of the starting materials. $HN(PPh_2NSiMe_3)_2$ is converted to the corresponding lithium salt by means of nBuLi or $LiNH_2$ and finally treated with Me_3SiCl to yield $Me_3SiNPPh_2NPPh_2N(SiMe_3)_2$ **8** [4].

Compound **11** is the first cyclophosphazene metal oxide [5]. It appears possible that **11** is dimeric in the solid state, since the only known cyclotriazaphosphazine metal oxide $[NPPh_2NC(4\text{-}CF_3C_6H_4)NReO_2]_2$ is also dimeric, which has been established by a single crystal structure investigation. The reaction of **11** with excess ArNCO (Ar = 2.6 diisopropylphenyl) leads to a mixture of products containing the cyclophosphazene metal imides **12** and **13**.

Compound **12** is generated by a [2+2] cycloaddition reaction. This addition at the ReN double bond of **11** leads to a distorted square-pyramidal arrangement of the ligands at the rhenium in **12**. An otherwise possible planarity of the six-membered ring is therefore removed. The average deviation from the plane is 20 pm.

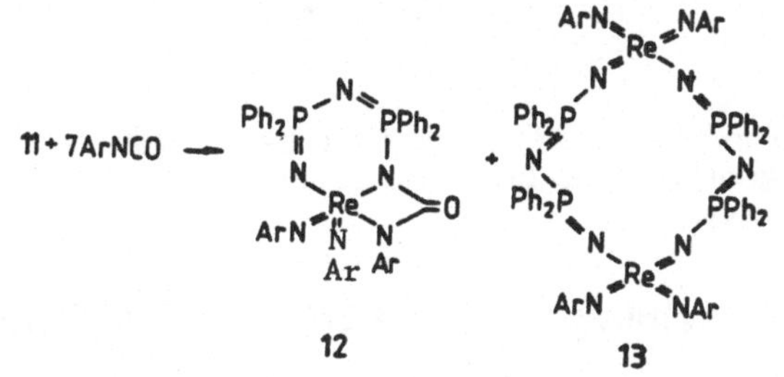

In the dimetallatetraimidophosphazene **13** which is only isolated in small quantities, the Re centers show an almost tetrahedral coordination and the twelve-membered ring is puckered. We assume that **13** is formed from the dimer **14** by opening two of the ReN bonds generating the twelve-membered ring **13**.

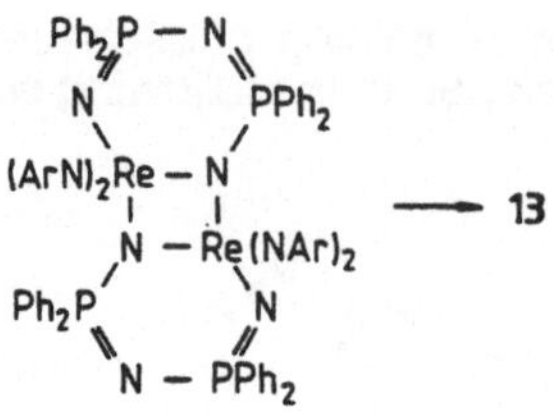

14

In the presence of Lewis acids the polymerization of $P_3N_3Cl_6$ leads to the corresponding polymer. Therefore, the formation of a phosphonium ion is assumed, which, as the rhenium atom in **14** initiates the cycloaddition. It is aparent that if the [2+2]-selfaddition is continued to result in larger molecules it could represent a novel polymerisation mechanism by cyclophosphazenes.

Compound $HN(PPh_2NSiMe_3)_2$ **10** has been used for the reaction with the metal alkyles $AlMe_3$, $GaMe_3$, $InMe_3$ or $ZnMe_2$ to yield for example **15** and **16** respectively under elimination of CH_4.

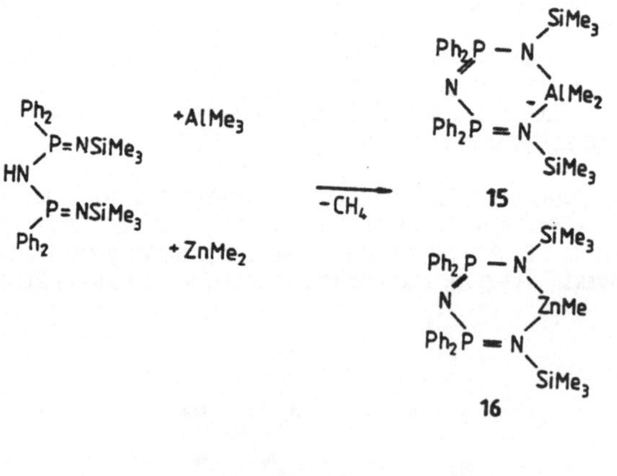

15

16

15 and **16** are crystalline white solids which have been characterized by single crystal X-ray structural analysis. In all cases we observed only the reaction of one methyl group at the metal center.

3 Compounds Containing Ti=N and Zr=N Double Bonds

Inter- and intramolecular cyclisation reactions are widely used in organic chemistry for synthesizing carbocycles. The tendency of a molecule with carbon-carbon multiple bonds to undergo a cycloaddition reaction with another unsaturated molecule depends on two factors, whether the other molecule contains isolated or conjugated double bonds, and whether the system is activated by heat or by light.

There is a considerable interest in the extent to which M=NR functional groups undergo such reactions. During our study of the chemistry of titanium-nitrogen compounds we observed that coordination of electron pair donors at the titanium atom leads to the stabilization of imidotitanium complexes [6,7]. Thus, reaction of the thiophospinic acid amide **17**

$$\mathrm{Ph_2P(S)-N(SiMe_3)_2} \xrightarrow[\text{2) py}]{\text{1) TiCl}_4} \mathrm{Ph_2P(S)-N=TiCl_2\cdot 3py}$$

17 **18**

with $TiCl_4$ and subsequent treatment of the initial product with pyridine (py) yields the orange red, crystalline compound **18**. In constrast, reaction of the oxygen analogue of **17** leads to an eight-membered ring compound [8]. The exchange of phenyl groups in **17** for isopropyl substituents gives the starting material **19**. The reaction of **19** with $TiCl_4$ leads in the presence of pyridine to the corresponding titanium imido complex [7], $(iPr)_2P(S)N{=}TiCl_2$ 3py **20**. However, the reaction of **19** with $TiCl_4$ in the presence of acetonitrile forms the [2+2] cycloaddition product **21** which can be converted to **20** in the presence of pyridine. The cycloaddition product **21** consists of a planar

```
        Me
        C
        N    iPr
  Cl\   |     |
     Ti  -   N  - iPr
  Cl/   |     |   /Cl
iPr - N  -  Ti
        |     |   \Cl
       iPr    N
              C
              Me
```

21

four-membered Ti_2N_2 ring with pairwise nonequal Ti-N bond lengths of 186.3(2) and 206.0(2) pm, respectively. Thus, the monomer-dimer formation can be directed by changing the basicity of the solvents. The imido ligand in **18** and **20** is bonded almost linearly to the metal [**18** Ti-N-P 171.4(2)°, **20** Ti-N-P 172.5(2)°].

A Cp* (Cp* = C_5Me_5) substituted titanium compound **22** containing a TiN double bond was obtained according to the following reaction sequence [9].

$$\mathrm{Cp^*TiCl_3 + Me_3SnNHtBu \longrightarrow Cp^*TiCl_2NHtBu + Me_3SnCl}$$

$$\mathrm{Cp^*TiCl_2NHtBu + LiN(SiMe_3)_2{\cdot}Et_2O + py \longrightarrow}$$
$$\mathrm{Cp^*Ti(Cl){=}NtBu{\cdot}py + LiCl + HN(SiMe_3)_2 + Et_2O}$$

22

py = pyridine

Similar to **22** $Me_3SiC_5H_4Ti(Cl){=}NtBu{\cdot}py$ **23** was prepared. Up to date a few other monomeric titanium imido complexes have been crystallographically characterized. All compounds knwon are summarized in Table 1.

Table 1. Monomeric titanium imido complexes

Compound		TiN bond length [pm]	Literature
$Ph_2P(S)N{=}TiCl_2{\cdot}3py$	**18**	172.0(2)	[6]
$(iPr)_2P(S)N{=}TiCl_2{\cdot}3py$	**20**	172.3(2)	[7]
$PhN{=}Ti(2.6\text{-}iPr_2C_6H_3O)_2{\cdot}2py'$		171.9(3)	[10]
$tBuN{=}Ti(Cl)C_5Me_5\ py$	**22**	169.8(4)	[9]
$(Et_4C_4NC_6H_4\text{-}NC_6H_4\text{-}NC_6H_4)Ti(OAr)_2$		170.8(5)	[11]
$tBuN{=}TiCl_2(OPPh_3)_2$		167.2(7)	[12]

py = pyridine, py' = 4-pyrrolidinopyridine

Wolczanski reported on the exposure of the alkyl complexes $(tBu_3Si)(THF)RTi{=}NSitBu_3$ (R = Me, tBu) to hydrogen in benzene for 3 h at 65°C leading to the formation of $[(tBu_3SiNH)Ti]_2(\mu\text{-}NSitBu_3)_2$ **24** and concomitant methane and isobutane, respectively. A single crystal X-ray structural determination confirmed a short Ti-Ti bond distance (Ti-Ti 244.2(1) pm). The μ-$NSitBu_3$ groups form asymmetric bridges and the bonding properties are best described by the following resonance forms.

$SitBu_3$ / N / $tBu_3SiN(H)\text{-}Ti — Ti\text{-}N(H)SitBu_3$ / N / $SitBu_3$ ⟶ $SitBu_3$ / N / $tBu_3N(H)\text{-}Ti — Ti\text{-}N(H)SitBu_3$ / N / $SitBu_3$

24 **24**

The first structural characterized zirconium imido complex of composition $Cp_2Zr = NtBu{\cdot}THF$ **25** was prepared in 1988 by Bergman et al. [14]. The starting material can be generated by heating Cp_2ZrMe_2 with one equivalent of $tBuNH_2$. This results in loss of one equivalent of methane and generation of the zirconocene methyl amide. Alternatively, the methyl amide can be prepared by treatment ofCpZrMe(Cl) with the lithium salt of $tBuNH_2$.

$$Cp_2Zr(Me)(NHtBu) \xrightarrow[\text{THF}]{\Delta} Cp_2Zr = NtBu{\cdot}THF$$

25

When the zirconozene methyl amide is subjected to thermolysis in THF compound **25** can be isolated.

The X-ray structural analysis of **25** showed a Zr-N bond length of 182.6(4) pm.

Rothwell et al. [15] reported on the structural characterization of [Zr(NC_6H_3iPr-2.6)($NHC_6H_3iPr_2$-2.6)$_2$·2py' **26** (py' = 4-pyrrolidinopyridine) and [Zr($OC_6H_3tBu_2$-2.6)(NPh)·2py' **27** with Zr-N bond lengths of 186.8(3) and 184.4(9) pm, respectively.

(η^5-C_5Me_5)Zr($NHC_6H_3iPr_2$-2.6)$_3$ **28** reacts at 85°C in the presence of pyridine under elimination of $H_2NC_6H_3iPr_2$-2.6 to yield ($\eta^5$$C_5Me_5$)Zr($NHC_6H_3iPr_2$-2.6)$NC_6H_3iPr_2$-2.6)·py **29** [16].

(η^5-C_5Me_5)Zr(NHR)$_3$ + py ——> (η^5-C_5Me_5Zr(NHR)(NR)·py + H_2NR

28 **29**

R = 2.6-$iPr_2C_6H_3$

The formation of **29** is based on ^{1}H-NMR and MS spectra. The monomeric structure of **29** was determined by a single crystal structure investigation. Compound **29** has a distorted piano-stool structure and contains three different Zr-N bonds. The bond length of the coordinating pyridine [234.5(6) pm], the amido Zr-N bond length [210.6(5) pm] and the imido bond distance [187.6(4) pm). The Zr=N-C bond angle is almost linear [171.4(4)°].

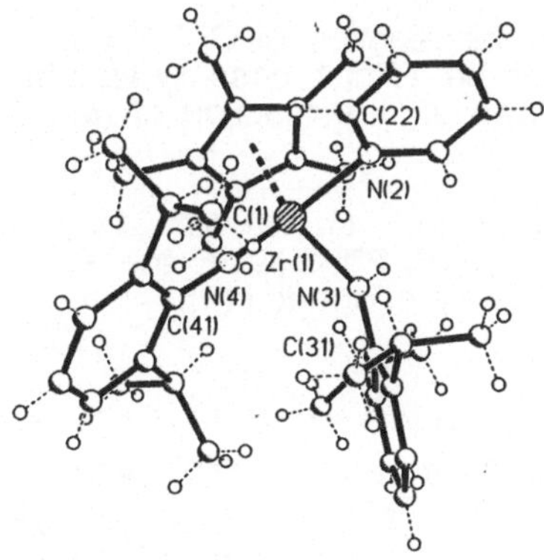

Figure 1 Molecular structure of **29**

4. Metallacyclosiloxanes

The phosphazene unit, -N=P(R_2)-, is isoelectronic with the siloxane group -O-Si(R_2)$_2$-. Consequently after preparing cyclometallaphosphazenes we were interested in synthesizing metallacyclosiloxanes. Our studies began with the reactions of (tBu)$_2$Si(OH)$_2$ **30** with $TiCl_4$, $TiBr_4$ and TiI_4 leading to the eight-membered ring compounds **31** [17].

$$2\ tBu_2Si(OH)_2 + 2\ TiX_4 \xrightarrow{-4HX} [tBu_2SiO_2TiX_2]_2$$

30 → 31

31a: X = Cl
31b: X = Br
31c: X = I

Compounds **31a** and **31b** were both investigated by X-ray diffraction. In both there is distorted tetrahedral geometry at the titanium atoms. The halides in **31a** may be replaced by different methods. When **31a** was reacted with CpNa(Cp = C_5H_5) the Cp substituted Ti compound **32** was isolated.

$$\mathbf{31a} + 2\ CpNa \xrightarrow{-2NaCl} [tBu_2SiO_2Ti(Cp)Cl]_2$$

32

Compound **32** is also obtained by the reaction of $CpTiCl_3$ with $(tBu)_2Si(OLi)_2$. Substitution at Ti in **31a** was found to be difficult, possibly as a result of the steric demands of the tBu_2Si groups. However, the reaction of $(Me_3Si)_2NLi$ leads to $[(Me_3Si)_2NClTiOSitBu_2O]_2$ **33**.

Attempts to react Cp^*M(Cp^* = C_5Me_5, M = Li, Na) with **31a** were unsuccessful. However, when Cp^*TiCl_3 was treated with $Ph_2Si(OLi)_2$ the eight-membered ring **34** as well as the six-membered ring **35** were formed [18].

$$Cp^*TiCl_3 + Ph_2Si(OLi)_2 \longrightarrow [Ph_2SiO_2Ti(Cp^*)Cl]_2 + Ph_2Si(O)_2[Ti(Cp^*)Cl]_2O$$

34 35

The halide-free eight-membered ring **36** is obtained by the reaction of **30** with $Ti(NEt_2)_4$.

$$2\,tBu_2Si(OH)_2 + 2\,Ti(NEt_2)_4 \xrightarrow{-4\,HNEt_2} [tBu_2Si(\mu\text{-}O)_2Ti(NEt_2)_2]_2$$

30 **36**

Of particular interest is compound **37**. It is formed according to the following equations.

$$Cp^{*}TiCl_3 + tBu_2Si(OH)_2 \longrightarrow Cp^{*}TiCl_2\text{-}OSitBu_2OH$$

30 **38**

$$2\ \mathbf{38} + 2\,H_2O \xrightarrow[-4\,HCl]{Et_3N} [tBu_2Si(\mu\text{-}O)_2Ti(Cp^{*})(OH)]_2$$

37

The structure of **37** was determined by X-ray diffraction. Compound **37** contains one O-H···O bond and one non-bridging OH group. The Ti-O bond length with the additonal hydrogen bond is longer [Ti-O 187.2(2) pm] in comparison to the other exocyclic Ti-O bond length [Ti-O 183.7(2) pm]. Thus, compound **37** demonstrates a frozen position of eliminating a water molecule. A side view of the molecular structure in the crystal is shown.

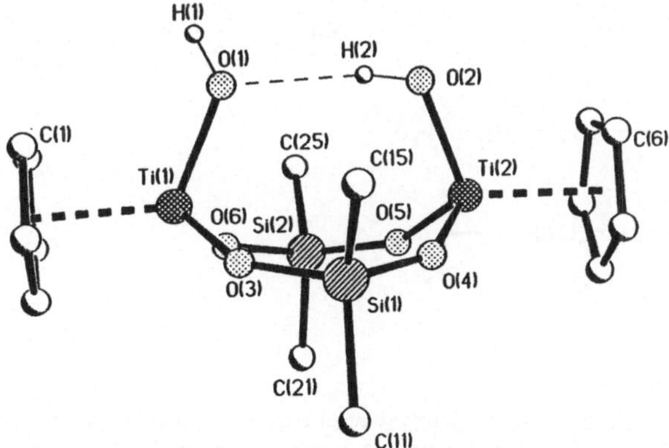

Figure 2 Side view of the molecular structure of **37**. The methyl groups are omitted for clarity

two molecules of THF as additional ligands, while the boron has a distorted trigonal planar coordination sphere [17,20].

Re_2O_7 and **30** yield the acyclic $tBu_2Si(OReO_3)_3$ **38** under elimination of water, while $TeCl_4$ gives the chloro-bridged $tBu_2Si(OTeCl_3)_2$ **34** with pentacoordinated tellurium atoms [21].

$$tBu_2Si(OH)_2 \ (\mathbf{30}) + Re_2O_7 \xrightarrow{-H_2O} tBu_2Si(O-ReO_3)_2 \ (\mathbf{38})$$

$$tBu_2Si(OH)_2 \ (\mathbf{30}) + 2\,TeCl_4 \xrightarrow{-2HCl} \mathbf{39}$$

In the case of $SnCl_4$ the formation of an eight-membered ring was not observed. The reaction proceeds according to the following scheme under formation of the adduct **40**.

$$2\,tBu_2Si(OH)_2 \ (\mathbf{30}) + 2\,SnCl_4 \longrightarrow tBu_2Sn \cdot [HO(H)O{-}SnCl_3]_2 \ (\mathbf{40}) + (tBu_2SiOH)_2O \ (\mathbf{41})$$

$$2\ \mathbf{30} \longrightarrow (tBu_2SiOH)_2O \ (\mathbf{41}) + H_2O$$

$$\mathbf{30} + 2\,[SnCl_3OH] \ (\mathbf{42}) \longrightarrow \mathbf{40}$$

This reaction can be interpreted in such a way that the silandiole **30** is converted to the disiloxandiole **41** under formation of water. Obviosly this water hydrolizes partially $SnCl_4$ forming $[SnCl_3OH]$ **42** [20].

This reaction can be interpreted in such a way that the silandiole **30** is converted to the disiloxandiole **41** under formation of water. Obviosly this water hydrolizes partially $SnCl_4$ forming $[SnCl_3OH]$ **42** [20].

The compounds $[tBu_2Si(O_2)MoO_2]_2$ **43** and $[tBu_2Si(O_2)V(O)Cl_2]_3$ **44** have been prepared using **30** and $VOCl_3$ or MoO_2Br_2, respectively [22]. While **43** forms an eight-membered ring **44** was isolated as a twelve-membered ring compound. In **44** the ring is puckered and the Si-O-V angels are varying between 141.7(2) and 173.8(3)°. These great differences are frequently observed in Si-O-metal systems.

3 tBuSi(OH)2 + 3 VOCl3 → 44

44

5. Model compounds for metal oxides on silica surfaces

Heterogeneous silica supported transition metal oxides play an important role as catalysts in industry [23,24]. These catalysts have increased the interest in the chemical processes which occur on the surface of heterogeneous catalysts. But to a large extent the processes taking place on the catalyst surface are not understood. Their study is hindered by the complicated structure of these silicate surfaces making the synthesis of model substances essential. Compounds **38** and **43** are models for silica supported metal catalysts which structurally resemble transition metal containing surface sites.

The substitution of two oxygen atoms in **38** occurs under elimination of CO_2 when this compound is treated with excess of 2.6-diisopropylphenylisocyanate (ArNCO). However, the compound expected is not the monomeric derivate **45** but rather the cluster **46**.

tBu2Si(OReO3)2 + 2 ArNCO → tBu2Si(OReO2=NAr)2 (not formed) / → 46

38 **45** **46**

In contrast to the starting material **38** where a distorted tetrahedral surrounding of the rhenium atoms is observed, in **46** a distorted octahedral geometry is found. The replacement of oxygen by the isolobal NR group changes the coordination number at rhenium from four to six. Under this observation we are assuming that the energy differences between tetrahedral and octahedral geometries are small and might be responsible for the catalytic properties of rhenium(VII) [25].

An interesting starting material for generating model compounds for metal oxides on silica surfaces is $tBuSi(OH)_3$ **47**. This compound was prepared by hydrolysis from $tBuSiCl_3$ and water in the presence of aniline in 94% yield.

$$tBuSiCl_3 + 3H_2O \xrightarrow[-3HCl]{} tBuSi(OH)_3$$

47

The X-ray crystal structure revealed that molecules of **47** are linked by hydrogen bridges to form corrugated layers. Bonding interactions between the layers are prevented by the hydrophobic tBu groups [26].

The reaction of **47** with Re_2O_7 in a 2:1 molar ratio affords $[tBuSiO(ReO_4)]_n$ (n = 3,4) after elimination of water and exchange of H for ReO_3. Recrystallization of the initial precipitate from boiling acetonitrile gave transparent crystals of $[tBuSiO(ReO_4)]_4$ **48**.

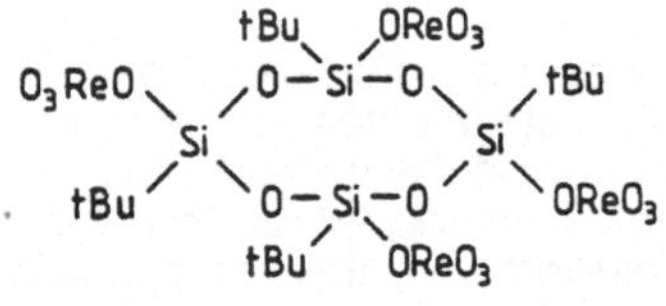

48

Compound **48** is the first example of a molecule containing four ReO_4 groups. In the crystal structure of **48** we observed that in the solid state all ReO_4 groups are on one side of the molecule.

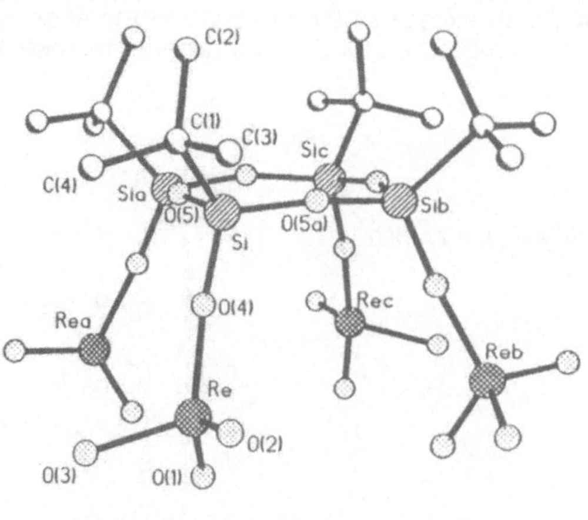

Figure 3 **Molecular structure of 48 in the crystalline state**

An introduction of two ReO_4 groups at antimony is possible when triphenylstibinoxide is treated with Re_2O_7 [27].

$$(Ph_3SbO)_2 + 2Re_2O_7 \longrightarrow 2Ph_3Sb(OReO_3)_2$$

49

In compound **49** antimony has a trigonal bipyramidal geometry where the phenyl groups are in equatorial and the ReO_4 groups in axial positions.

5. Cyclometallaborazines

Borazine was first reported by Stock and Pohland [2]. It is also known as inorganic benzene and was isolated from the mixture of products obtained by reacting B_2H_6 and NH_3. Borazine has a resemblance to benzene and the physical properties of the two compounds are similar. The planarity of the borazine molecule is shown by MO calculations to be stabilized by the π bonding, however, the π electrons are only partially delocalized. The number of borazines substituted at the ring framework is legion. Only recently the first examples were reported which contain transition metal atoms as building blocks in the borazine framework.

Methyl-bis(methyl(trimethylsilyl)aminophenylboryl)amine **50** served as educt for the synthesis of the target molecule.

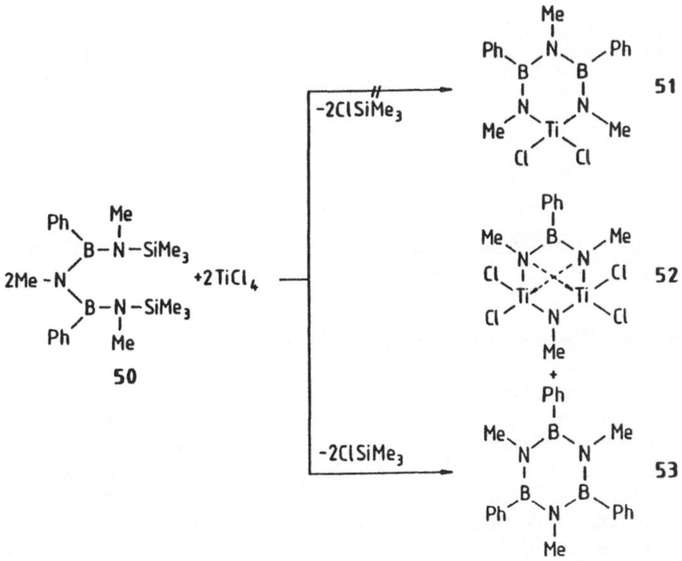

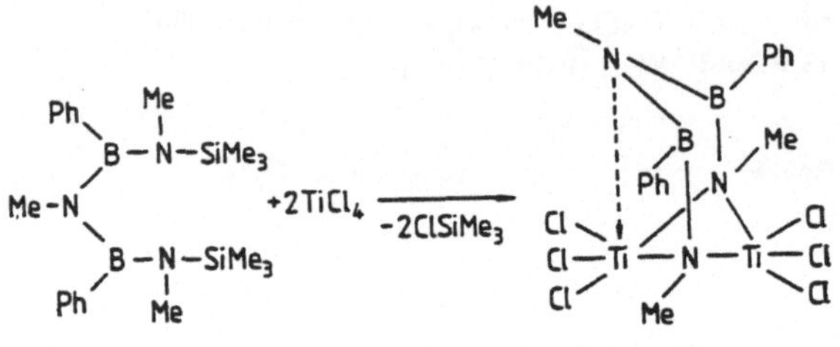

Compound **54** decomposes via an S_{Ni} reaction to give **52** and $PhBCl_2$. The latter compound then reacts with a further equivalent of **50** to give **53**. The structure of **52** can be described as a geometric body whose surface consists of four bent irregular squares. The reason for this nonplanar configuration are the strong intramolecular Ti···N interactions [28].

The first six-membered borazine containing a tellurium atom as a building block in the ring frame-work was obtained from **50** and $TeCl_4$. Compound **55** was investigated by an X-ray structural analysis showing the six-membered ring in a non planar configuration [29].

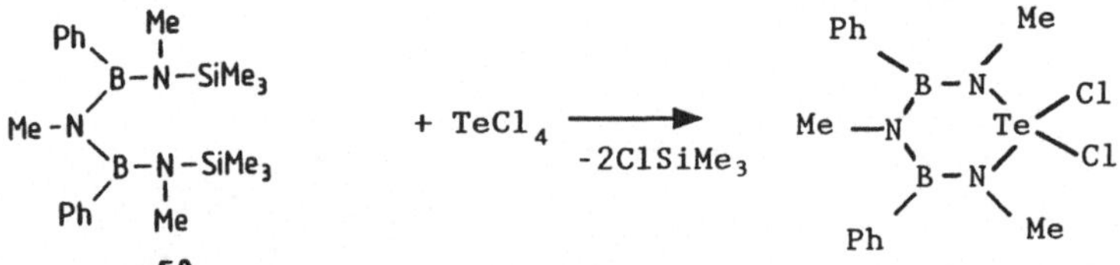

In addition it is worth mentioning that the number of metal containing aromatic systems are rare [30-33].

6. Organometallic precursors for CVD

Aluminum microstructures can be generated efficiently using thermal CVD from (trimethylamine)trihydridoaluminum **56** on laser generated spatially selective prenucleation pattern of a Pd-catalyst. The aluminum structure generated at 200°C was 50 μm wide and 7 μm high. The two step process combines the advantages of two methods. The laser induces the spatially selective pattern of the Pd-catalyst, whereas the macroscopic growth is done by conventional well-understood chemical vapor deposition [34].

AlN is generated by CVD using $[Mes_2AlNH_2]_2$ **57**(Mes = Mesityl) as a starting material [35]. Compound **57** is prepared from Mes_3Al and ammonia resulting in the formation of the adduct $Mes_3Al{\cdot}NH_3$ which eliminates at elevated temperatures MesH to yield **57**. The advantage of compound **57** for this process is prenucleation pattern of a Pd-catalyst. The aluminum structure generated at 200°C was 50 µm wide and 7 µm high. The two step process combines the advantages of two methods. The laser induces the spatially selective pattern of the Pd-catalyst, whereas the macroscopic growth is done by conventional well-understood chemical vapor deposition [34].

AlN is generated by CVD using $[Mes_2AlNH_2]_2$ **57**(Mes = Mesityl) as a starting material [35]. Compound **57** is prepared from Mes_3Al and ammonia resulting in the formation of the adduct $Mes_3Al{\cdot}NH_3$ which eliminates at elevated temperatures MesH to yield **57**. The advantage of compound **57** for this process is its remarkable stability towards traces of moisture compared to alkyl derivatives of aluminum [35].

H H
Mes N Mes
Al Al
Mes N Mes
H H

57

The generation of pure InP using $[(Me_3SiCH_2)_2InPHtBu]_2$ **58** and a 514.5 nm Ar-ion-laser for pyrolysis was not successful. Besides InP the sample contained a high percentage of SiC [36]. Another precursor was prepared from $(Me_3SiCH_2)_3In$ and $AdPH_2$ (Ad = adamantyl) in the presence of $AgNO_3$ leading to $[(Me_3SiCH_2)_2InPHAd]_2$ **59** in 30% yield [37]. Pyrolysis experiments have so far not been accomplished.

Acknowledgement

The author greatly appreciates the contributions of his students and colleagues whose names appear in the references. Financial support by the Deutsche Forschungsgemeinschaft, the Fonds der Chemischen Industrie and the VW-foundation is highly acknowledged.

References

[1] H.W. Roesky, Synlett, 1(1990) 651, and literature quoted herein.

[2] A. Stock, E. Pohland, Chem. Ber. 59 (1926) 2215.

[3] H.W. Roesky, K.V. Katti, U. Seseke, M. Witt, E. Egert, R. Herbst, G.M. Sheldrick, Angew. Chem. 88 (1986) 447; Angew. Chem. Int. Ed. Engl. 25 (1986) 477.

[4] R. Hasselbring, H.W. Roesky, M. Rietzel, M. Noltemeyer, Phosphorus, Sulfur, Silicon, in press.

[5] R. Hasselbring, H.W. Roesky, M. Noltemeyer, Angew. Chem. in press.

[6] H.W. Roesky, H. Voelker, M. Witt, M. Noltemeyer, Angew. Chem. 102 (1990) 712; Angew. Chem. Int. Ed. Engl. 29 (1990) 669.

[7] H.W. Roesky, T. Raubold, M.Witt, R. Bohra, M. Noltemeyer, Chem. Ber. 124 (1991) 1521.

[8] M. Witt, H.W. Roesky, D. Stalke, T. Henkel, G.M. Sheldrick, J. Chem. Soc. Dalton Trans 1991, 663.

[9] Y. Bai, M. Noltemeyer, H.W. Roesky, Z. Naturforsch. B 46 (1991) 1357.

[10] J.E. Hill. R.D. Profilet, P.E. Fanwick, I.P. Rothwell, Angew. Chem. 102 (1990) 713; Angew. Chem. Int. Ed. Engl. 29 (1990) 664.
[11] J.E. Hill, P.E. Fanwick, I.P. Rothwell, Inorg. Chem. 30 (1991) 1143.
[12] C.H. Winter, P.H. Sheridan, T.S. Lewkebandara, M.J. Heeg, J.W. Proscia, J. Am. Chem. Soc. 114 (1992) 1095.
[13] C.C. Cummins, C.P. Schaller, G.D. Van Duyne, P.T. Wolczanski, A.W.E. Chan, R. Hoffmann, J. Am. Chem. Soc. 113 (1991) 2985.
[14] P.J. Salsh, F.J. Hollander, R.G. Bergman, J. Am. Chem. Soc. 110 (1988) 8729.
[15] R.D. Profilet, C.H. Zambrano, P.E. Fanwick, J. J. Nash, I.P. Rothwell, Inorg. Chem. 29 (1990) 4364.
[16] Y. Bai, H.W. Roesky, M. Noltemeyer, M. Witt, Chem. Ber. 125 (1992) 825.
[17] A. Haoudi-Mazzah, A. Mazzah, H.-G. Schmidt, M. Noltemeyer, H.W. Roesky, Z. Naturforsch. 46B (1991) 587.
[18] F.Liu, H.-G. Schmidt, M. Noltemeyer, C. Freire-Erdbrügger, G.M. Sheldrick, H.W. Roesky, Z. Naturforsch. in press.
[19] F. Liu, H.W. Roesky, H.-G. Schmidt, M. Noltemeyer, Angew. Chem. in preparation.
[20] A. Mazzah, A. Haoudi-Mazzah, M. Noltemeyer, H.W. Roesky, Z. Anorg. Allg. Chem. 604 (1991) 93.
[21] H.W. Roesky, A. Mazzah, D. Hesse, M. Noltemeyer, Chem. Ber. 124 (1991) 519.
[22] H.-J. Gosink, H.W. Roesky, M. Noltemeyer, H.-G. Schmidt, C. Freire-Erdbrügger, G.M. Sheldrick, Chem. Ber. in preparation
[23] F.J. Feher, J. Am. Chem. Soc. 108 (1986) 3850.
[24] F.J. Feher, D.A. Newman, J.F. Walzer, J. Am. Chem. Soc. 111 (1989) 1741.
[25] H.W. Roesky, D. Hesse, R. Bohra, M. Noltemeyer, Chem. Ber. 124 (1991) 1913.
[26] N. Winkhofer, H.W. Roesky, M. Noltemeyer, W.T. Robinson, Angew. Chem. in press.
[27] U. Wirringa, H.W. Roesky, H.-G. Schmidt, M. Noltemeyer, Chem. Ber. in preparation.
[28] H.-J. Koch, H.W. Roesky, R. Bohra, M. Noltemeyer, H.-G. Schmidt, Angew. Chem. in press.
[29] H.-J. Koch, H.W. Roesky, unpublished results.
[30] G.P. Elliott, W.R. Roper, J.M. Waters, J. Chem. Soc. Chem. Commun. 1982, 811.
[31] M.S. Kralik, A.L. Rheingold, R.D. Ernst, Organometallics, 6 (1987) 4118.
[32] J.R. Bleeke, Y.-F. Xie, W.-J. Peng, M.Y. Chiang, J. Am. Chem. Soc. 111(1989) 4118.
[33] J.R. Bleeke, Y.-F. Xie, L. Bass, M.Y. Chiang, J. Am. Chem. Soc. 113 (1991) 4703.
[34] O. Gottsleben, H.W. Roesky, M. Stuke, Adv. Materials 3 (1991) 201.
[35] T. Belgardt, W. Rockensüß, H.W. Roesky, unpublished results.
[36] U. Dembowski, M. Noltemeyer, W. Rockensüß, M. Stuke, H.W. Roesky, Chem. Ber. 123 (1990) 2335.
[37] U. Dembowski, H.W. Roesky, E. Pohl, R. Herbst-Irmer, D. Stalke, G.M. Sheldrick, Z. Anorg. Allg. Chem. in press.

METHYLIDENETITANACYCLOBUTANE vs. TITANOCENE-VINYLIDENE - VERSATILE BUILDING BLOCKS

Rüdiger Beckhaus

Institute of Inorganic Chemistry, Technical University of Aachen, Prof.-Pirlet-Strasse 1, W-5100 Aachen, FRG

Summary

The reaction of vinyllithium with $Cp^*_2TiCl_2$ ($Cp^*=[C_5(CH_3)_5]$) in a molar ratio of 2:1 quantitatively yields the unexpected titanacyclobutane derivative $Cp^*_2\overline{TiC(=CH_2)CH_2CH_2}$ **2**. The isolated complex **2** has been characterized by spectroscopic methods and by X-ray structure determination. The chemical reactivity of **2** is characterized by the splitting of the four membered ring at higher temperatures, forming a titanocene fragment $\{Cp^*_2Ti=C=CH_2\}$ (**6**) and ethylene. This cycloreversion is confirmed by the fragmentation behavior in the mass spectrometer, and some aspects of it have been studied by Extended Hückel Model calculation. The system titanacyclobutane **2** and titanocenevinylidene **6** can be used in a large scale of different organometallic reactions. So **2** exhibits typical properties of an alkyl complex and undergoes insertion reactions with isonitriles or ketenes, forming 5- and 6-membered metallacycles. On the other hand cycloreversion products can be prepared due to the typical vinylidene character of **6**. By reaction of **6** with metal carbonyls, followed by a vinylidene-acetylene rearrangement, 5-membered cyclic FISCHER-Carbene complexes can be isolated in high yield.

1 Introduction

Early transition-metal complexes have become useful reagents for the investigation of many important reactions in organometallic chemistry, including olefin metathesis [1, 2] and polymerization [3-5] , homogeneous reduction of dinitrogen [6-9] and carbon monoxide [10,

11], hydrozirconation- [12, 13] in combination with transmetallation reactions [14, 15] and at last the activation of C-H bonds [16-19]. The diverse reactivity exhibited by these complexes in many instances relies on the availability of a vacant orbital at the d^0 metal center. The orientation of this orbital in the coordinatively unsaturated $(Cp)_2$- or $(Cp^*)_2ML_2$-type complexes, as shown by various theoretical [20-23] and experimental [10, 24-26] works, is directed primarily normal to the plane that bisects the L-M-L bond angle.

Vinyl compounds of electron poor transition metals

During the past decade metal bounded vinylgroups have developed a new area of an important class of organometallic reagents. In particular in the case of electron poor metals, such as titanium, zirconium and hafnium, it is a very reactive, showing typical subsequent reactions of these substances [27]. So diene complexes (**1a,b**) are formed immediately even at low temperatures, if the metallocene chlorides of zirconium and hafnium (Cp_2MCl_2, Cp: C_5H_5, M: Zr, Hf) react with vinyllithium in a molar ratio of 1:2. Vinyl transition-metal compounds could not be detected as intermediates [28, 29]. On the other hand using substituted compounds of vinyllithium or by reactions of $Cp^*_2MCl_2$ (M = Ti, Zr, Hf) with vinyllithium in a molar ratio 1:2 the corresponding divinyl compounds are formed. The hafnium compound can be isolated pure. Titanium and zirconium compounds can be detected by low-temperature NMR spectroscopy, but at they undergo rearrangements at room temperature. Starting from $Cp^*_2Zr(CH{=}CH_2)_2$ the metallacyclopentene **3** is formed via ß-H-elimination and formation of a zirconocene(acetylene)(ethylene)-intermediate **5** [30, 31]. Surprisingly, $Cp^*_2Ti(CH{=}CH_2)_2$ is converted quantitatively into the metallacyclobutane **2** exhibiting a typical metallacyclobutane structure and reactivity [30, 32].

The use of diene complexes [33, 34] respectively metallacyclopentene precursors [35-38] as selective reagents in organic synthesis is well known. Especially the quantitative and selective formation of the titaniumcyclobutane **2** via vinyl- and vinylidenintermediates **4** is a very simple route to this attractive class of molecules.

Scheme 1

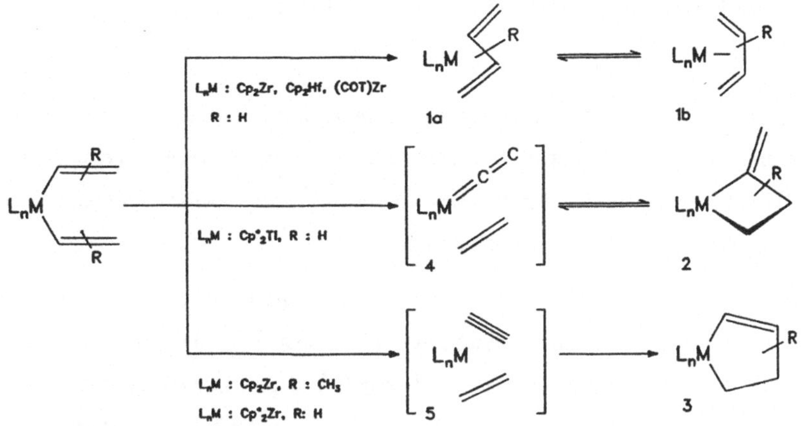

L_nM : Cp_2Zr, Cp_2Hf, (COT)Zr
R : H

L_nM : Cp^*_2Ti, R : H

L_nM : Cp_2Zr, R : CH_3
L_nM : Cp^*_2Zr, R: H

1a 1b 4 2 5 3

Metallacyclobutanes of electron poor metals

Titanocene metallacyclobutanes [39-41] show a wide variety of reactivities with organic and inorganic reagents[42-45]. Their reactions include methylene transfer to organic carbonyls [46-48], formation of enolates [49-51], electron transfer from activated alkyl chlorides [49], olefin metathesis [52, 53], ring-opening polymerization [54, 55], and complexation of metal halides [56-58]. All these reactions presumably occur through a reactive intermediate that exhibits behavior consistent with that of transition-metal carbenes [59, 60]. The intermediate has been postulated to bee a free titanocene methylidene or a titanocene methylidene olefin complex [61]. The cleavage of the metallacycle is the rate-determining step in these reactions [62, 63].

Metallacyclobutanes have been prepared by the reaction of Cp_2TiCl_2 with di-GRIGNARD reagents of 1,3-dibromopropanes [41, 64], starting from the TEBBE-reagent $Cp_2\overline{TiCH_2Al(Me)_2Cl}$ [52], or by using aryne-olefine-metallocene intermediates and phosporylides [65, 66].

In comparison to these "classical" metallacyclobutanes - as precursors of methylidene derivatives - the unconventional methylidenemetallacyclobutane **2** exhibiting an exocyclic methylidene group, has turned out to be a good starting material for species regarded as metalla-allene or titanocenevinylidene intermediates. Preparation of **2** follows in a simple manner to the "vinyl route".

Vinylidene complexes of electron poor transition metals

The chemistry of vinylidene $H_2C=C$:, as the most simple unsaturated carbene [67, 68] and especially of vinylidene transition-metal complexes has attracted a great deal of attention [69-71]. Terminal metal vinylidene complexes are species containing the formal metal-carbon-carbon cumulene bond system.

There are, however, only a few examples of vinylidene complexes of electron poor metals. Titaniumvinylidene derivatives have been of particular value in organic synthesis [42, 47, 72, 73]. Vinylidene derivatives of transition metals are discussed as popular models for the chain-lengthening/homologation steps in the methylene polymerization in the FISCHER-TROPSCH synthesis of hydrocarbons [74-76], especially in the case of tantalocene derivatives [77, 78].

Several methods have been employed for the preparation of mononuclear vinylidene complexes: 1. from 1-alkynes via a formal 1,2-hydrogen shift [79-82], 2. by addition of electrophiles to metal alkinyl derivatives [83-87], 3. by deprotonation of carbyne complexes [88, 89], 4. by formal dehydration of acyl complexes [90-93] and 5. starting from vinyl- [30, 32, 77, 78, 94, 95] or other vinylidene derivatives [96].

2 Structure and Reactivity of Bis(π-pentamethylcyclopentadienyl)-(2-methylidene)titanacyclobutane

Bis(π-pentamethylcyclopentadienyl)titanium dichloride, $Cp^*_2TiCl_2$ reacts with vinyllithium in a molar ratio of 1:2 to form the corresponding divinyl compound $Cp^*_2Ti(CH=CH_2)_2$ at low temperatures (-20 - -10 °C). By heating up to room temperature the metallacyclobutane 2 is formed and can be isolated in form of orange red crystals in nearly quantitative yield. The formation of 2 is indicative of a transient vinylidene-ethylene species 4 (Scheme 2) formed after vinylic α-hydrogen transformation from one vinylgroup to the other.

Schema 2

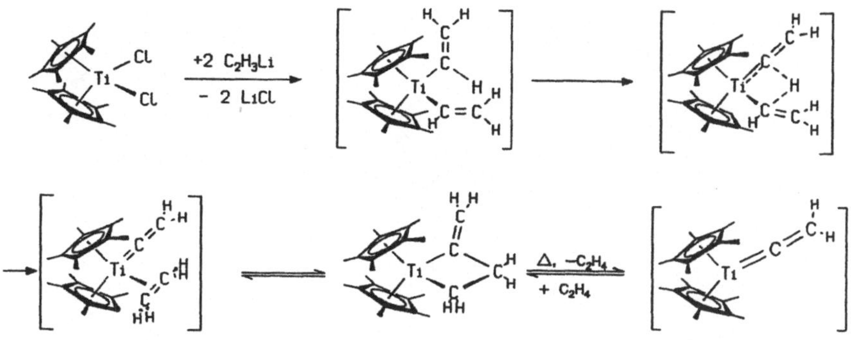

4 2 6

The titanacyclobutane complex **2** crystallizes in form of red needles and is thermally stable up to 170 °C in the solid state. Furthermore it is possible to handle it for some minutes in the air. The structure of the titanacyclobutane ring was confirmed by X-ray diffraction. Crystals, suitable for X-ray structure determination, were obtained by slow cooling of a small amount of a solution in pentane. The titanium atom is coordinated in a pseudotetrahedral manner, with typical distances to the Cp*-ligands of 2.120 and 2.125 Å and a Cp*-Ti-Cp* angle Φ = 138.5°. This is likely to be a charcteristicle feature of Titanocene(IV)-complexes ($Cp^*_2TiCl_2$,Ti-Cp* 2.13 Å, Φ = 137.4°, [97])), whereas for Titanocen(II)-complexes as $Cp^*_2Ti(C_2H_4)$ larger Φ-angles and shorter Cp*-Ti bond length are found (Ti-Cp*: 2.092 Å, Φ= 143.6°, [98]). The metallacyclic ring exhibits a nearly planar framework.

On the other hand, there are different bond lengths of the Ti-C bonds - one to the quaternary C-atom of 2.068 Å (Ti-C(03)) and one to the Csp3-atom of 2.137 Å (Ti-C(01)). The C-C distances between C(02) and C(01), C(03) respectively, are similar (1.520(10), 1.521(10) Å). The exocyclic C-C double bond length is in a normal range. That means that the complex exists in the solid state at room temperature as a metallacyclic ring. The alternative open titanocenevinylidene-ethylene structure **4**, as discussed for a titanocene-methylene-acetylene structure of a metallacyclobutene [99], is not formed at room temperature in the solid state. The value of the angle Ti-C(03)-C(04) of 152° is surprisingly high. Perhaps, it is due to a partial titanium-carbon double bond character, in connection with a streching of the four membered ring in the x,y-plane forming a relative short distance Ti-C(02).

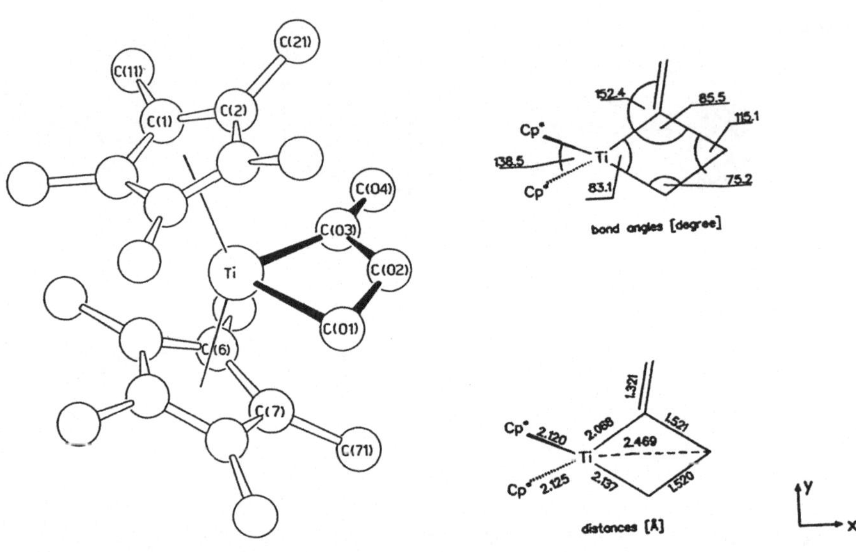

Figure 1. Perspective view of the molecular structure of $Cp^*_2\overline{TiC(=CH_2)CH_2CH_2}$ (2)

The chemical reactivity of the titanacyclobutane ring 2 is characterized on the one hand by typical properties of two different types of titanium-carbon σ-bonds and on the other hand by the splitting of the four membered ring at higher temperatures forming a titanocene vinylidene fragment 6 and ethylene. This cycloreversion is confirmed by the fragmentation behavior in the mass spectrometer and some aspects of it have been studied by extended Hückel calculations [32].

The mass spectrum clearly shows the molecular peak of 2 (372 m/z) and as a relatively stable intermediate the signals of the vinylidene fragment 6 (344 m/z). Finally the typical signals of the Cp^*_2Ti fragment are detectable. The transformation of the fragmentation behavior in the mass spectrometer into the preparative chemistry can be the entrance to vinylidenes, homologous derivatives of the well known methylidene fragments. From this point we have carried out Extended HÜCKEL calculations (calculated with Cp instead of Cp*) for systems 7-9 [32].

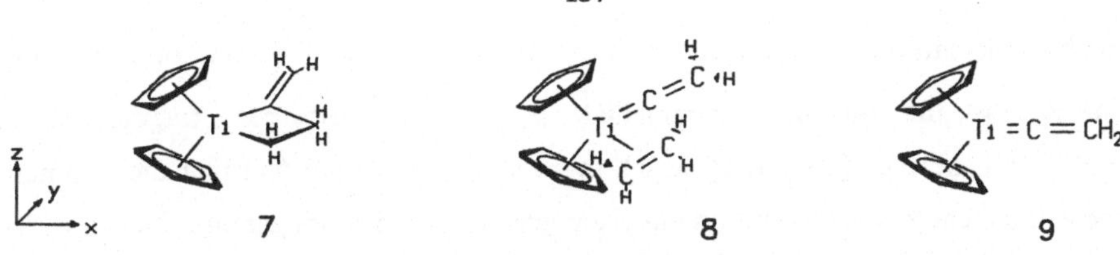

The minimum geometry of 7 agreed with the results of the structure determination by X-ray diffraction of 2 discussed before. Especially in the case of the angle Ti-C-C a value of 152° is found for the planar ring (calculated and measured). In the case of the ground state of the vinylidene-ethylene intermediate 8, a complex with two different π-acceptor properties, it was found, that the vinylidene group is bound to the metal centre much more strongly than the ethylene ligand. In detail that means that the way from the four membered ring to the vinylidene intermediate is a continuous one and allowed by symmetry. The alternative cycloreversion, forming titanocene methylidene and allene is unfavorable because of thermodynamic and kinetic reasons (figure 2). Even under experimental conditions, only the liberation of ethylene and the formation of metal vinylidene have been found.

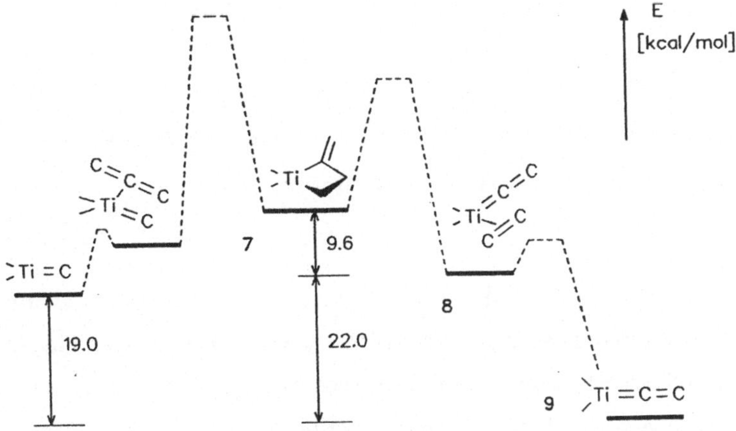

Figure 2. Schematic representation of the energy profile of the methylidenetitanacyclobutane-system

The occupied molecular orbitals of the titanocene vinylidene fragment are the π-orbital of the Ti-C double bond (HOMO) and the π-C-C-MO of the vinylidene group and at last the σ-orbital of the Ti-C bond. Owing to the character of the metal fragment orbitals b2 leads to a strong back-donation and a high barrier of rotation around the Ti-C bond of 32 kcal/mol. That means that the vinylidene intermediate is a real metal-carbene fragment of the SCHROCK-carbene type [100, 101]. Much more surprising is the fact that the symmetric vinylidene fragment 9 is not the minimum geometry. Indeed, the elongation of the vinylidene group of 35° from the central position stabilizes the molecule by 6.5 kcal/mol. That means the C_s-symmetry (**9a**) is prefered in the ground state rather than the C_{2v}-symmetry. Owing to the HOMO/LUMO gap of 1.98 eV, the vinylidene titanocene exists in a singlett ground state (figure 3).

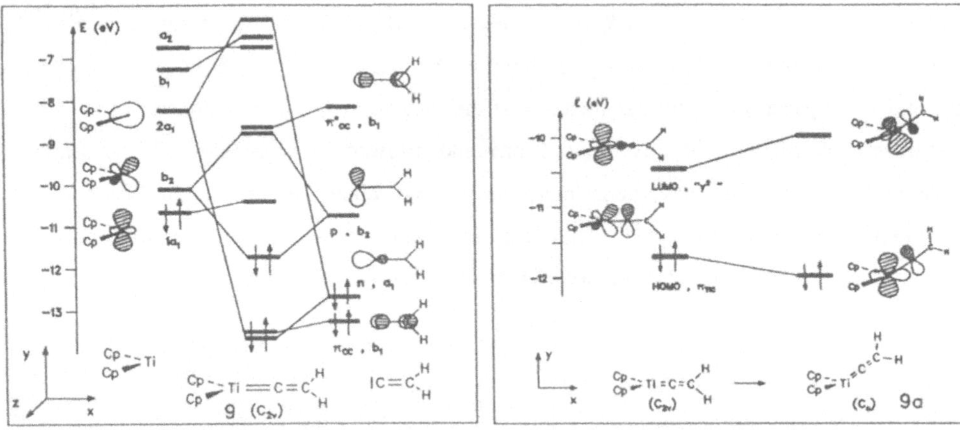

Figure 3. Orbital interaction diagram for a Cp_2Ti-fragment (C_{2v}) and a vinylidene fragment (C_{2v}) for trigonal planar $Cp_2Ti{=}C{=}CH_2$ 9 and for **9a** (C_s) (only relevant MO-orbitals)

Finally it must be mentioned that the LUMO of the vinylidene fragment is hybridized in the direction to the open coordination site. The HOMO is an excellent donor orbital to build up a backdonation with acceptor orbitals of ligands, such as ethylene. So we can summarize that, using the vinylidene intermediate, the symmetry of the π- and π^*-orbitals of the Ti-C-double bound is inverted, so that suprafacial cycloaddition reactions are allowed in the ground state.

For these reasons the titanocene vinylidene intermediate, generated by thermal reaction, should be a versatile reagent for cycloaddition reactions with a large scale of substrates, especially with oxygen-containing unsaturated substrates, such as ketones, ketenes or heterocumulenes, for instance metal carbonyls (chapter 4).

However, as well in the titanacyclobutane as in the titanocene vinylidene intermediate there are titanium-carbon bonds of different type, which can undergo insertion reactions (chapter 3).

3 Insertion Reactions

Carbon monoxide reacts with **2** at room temperature under normal pressure. As metal containing reaction product, however only the dicarbonyl derivative **10** is isolable. A monoinsertion product **11** has not been detected under these reaction conditions. But if isonitriles or ketenes are used under mild conditions (20 °C), insertion products are isolable. In the case of tert-butyl-, cyclohexyl- and (trimethylsilyl)methyl-isocyanide, dark green solutions are formed immediately, from which crystals of **12** can be isolated [102]. Characterization indicated a selective insertion in the Ti-$C_{sp}3$-bond under formation of five membered metallacyclic rings. Insertion reactions in the Ti-$C_{sp}2$-bond are not observed.

Scheme 3

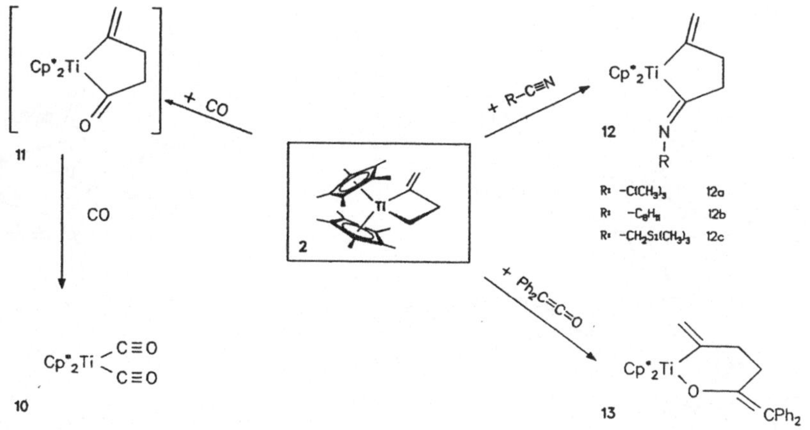

The molecular structure of **12b** was determined by single crystal structure determination. The ligands provide a pseudotetrahedral environment of the metal centre. The structural parameters about the Ti atom exhibit normal values. The respective Ti-C distances of Ti-C1 2.191(4) and Ti-C4 2.177(4) Å, the Ti-Cp* distances of 2.118(1) and 2.114(1) Å and the Cp*-Ti-Cp* angle of 141.1° are comparable to those observed for related Cp^*_2Ti-compounds containing four- [103, 104] and five-membered [105, 106] rings. Group 4 metallocene complexes with donor atoms (**D**) in the alkyl chain, like acyl (-C(=O)R), iminoacycl

(-C(=NR)R') [107] and alkoxymethyl (-CH_2OR) complexes [108-110] are known to adopt different bonding modes: η^2-outside (**A**), η^2-inside (**B**), or a η^1-outside type structure **C**.

A: η^2–outside B: η^2–inside C: η^1–outside

The complexes **12** exhibit an IR ν(C≡N) bands at 1564-1565 cm^{-1} (**a**,**b**) and a ^{13}C-NMR (C_6D_6, 50MHz) imino carbon resonance at 234.6 (**12a**), 236.3 (**12b**) respectively 240.1 ppm (**12c**). All values are typical for a (η^1-iminoacyl) complexes [107]. The iminoacyl functional group is clearly η^1-bonded to titanium due to the angle Ti-C1-N (123.5(3)°) and due to the Ti-N- (3.080(3) Å) and C1-N- (1.270(5) Å) distances. For a η^2-iminoacylstructure Ti-N and Ti-C bond lengths of 2.125 respectively 2.080 Å are found [111]. Double insertion products or other ring enlargements like in the case of zirconacyclobutanes [112, 113] are not observed.

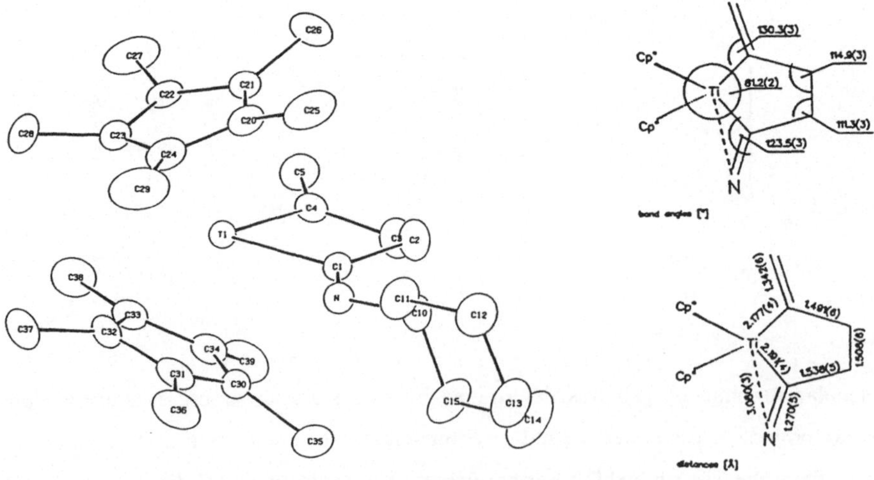

Figure 4. Perspective view of $Cp^*_2TiC(NC_6H_{11})CH_2CH_2C{=}CH_2$ **12b** with appropriate numbering scheme.

But on the other hand the six-membered metallacycle **13** is formed by insertion using diphenyl ketene and **2**. This complex is isolable in form of dark red airstable crystals [102].

The molecular structure of the metallacyclic compound **13** has been established by single-crystal X-ray analysis. A SCHAKAL view is shown in figure 5.

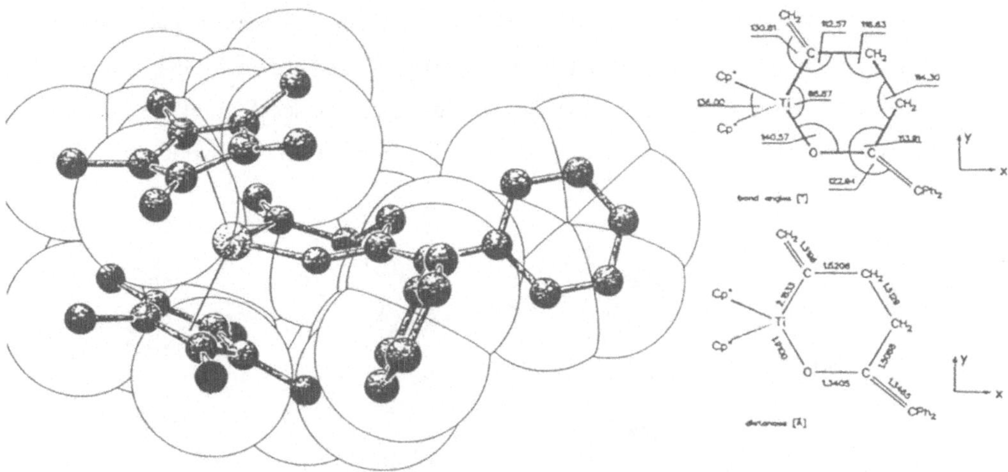

Figure 5. Schakal drawing and selected bond distances and angles of **13**.

Only a few examples of six-membered metallacyclic rings of the electron poor metals are known up to now. There is only one example of a six membered titanacycle in form of the carbonylation product of an oxametallacyclopentane derivative [114]. The Ti-C-bond distance in **13**, 2.1633 Å, corresponds to a single covalent Ti-C_{sp}2-bond, whereas the Ti-O-bond (1.9100 Å) is longer than in [114] (1.844 Å). The six-membered-ring is not planar, the titanium- and oxygen-atoms and the exocyclic methylidenegroup are nearly in plane, forming a pseudo envelope structure.

4 Cycloaddition Reactions

Under thermolytic conditions there is a selective liberation of ethylene generating the titanocene vinylidene fragment **6**. But this highly unsaturated species is very reactive, if there are no other acceptor ligands, a C-H-bond of one Cp* group forms a fulvene- (Fv), a so called "tuck-in" complex Cp*(Fv)TiCH=CH_2 **14** in form of dark green crystals, with a metal-bonded vinyl group [115]. In the NMR-spectra two different vinyl groups are detectable. For the isomer **14a**, main product (95%), a high field shift of the CH-proton of the vinyl group in connection with a very broad signal is observed. Perhaps it is due to a fast exchange of CH_2- and CH-protons of the vinyl- and the fulvene ligand respectively. The ^1H-NMR spectrum (270 MHz) shows two doublets with a coupling constants $^2J_{HH}$= 4.2 Hz assigned to the diasteriotopic CH_2-protons of the fulvene ligand. This coupling constant is more consistent with the geminal sp^2 protons of structure **A** (normaly 0-3 Hz) than with the geminal sp^3 hydrogens of structure **B** (normaly 12-15 Hz). The gated carbon spectrum of the CH_2 group resonates at 76.2 ppm with $^1J_{CH}$= 150 Hz, values most consistent with sp^2 hybridization of the carbon atom [116, 117].

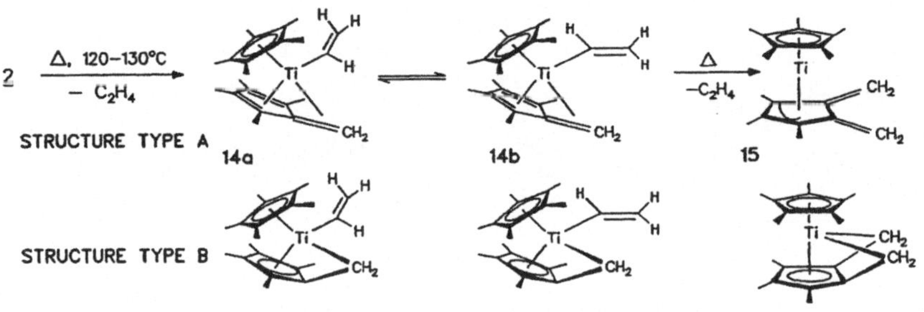

Prolonged heating of <u>**14**</u> at 130 °C (heptane solution) resulted in the formation of ethylene and a light blue organotitanium compound <u>**15**</u>.

The mass- and NMR-data are in complete accordance with the formulation <u>**15**</u> [115]. NMR data indicate that also structure **A** (sp^2 hybridized carbon) is the best representation. But we were unsuccessful to solve the structure by diffraction methods. It is hampered by disorder problems.

An unexpected formation of a vinyl titanocene derivative <u>**16**</u> occurs by heating <u>**2**</u> in the presence of acetone. In addition to the vinyl group an enolate fragment is present on the titanium centre. That means, instead of a cycloaddition reaction a protonation of the α-C-atom occurs due to the enol form of the acetone molecule ("en-reaction" [51, 50]). By using perdeuterated acetone, a deuterium atom was transformed into the α-position of the Ti=C=CH_2-group. With other enolizable ketones, the vinylidene fragment also formed metal enolates. The resulting (E)/(Z) ratio depends on the nature of R of the ketones [115]. With non-enolizable ketones, like in the case of Ta- [118], Zr- [119, 120] or other Ti-derivatives [121], metallaoxetanes are not observed under these reaction conditions (90°C, n-heptane).

ith aliphatic alcohols 2 formed vinyltitaniumalkoxides 17 in high yield [115] by heating up to 50 °C. At lower temperatures (20-50°C) no reaction occurs. The course of reaction is in accordance with a selective protonation of 6 in the α-postion, proven by using CH_3OD, like it is characteristic in the case of vinylidenes with nucleophilic α-C-atoms [122]. By using stoichiometric amounts of phenols 18 is formed by a selective splitting of the Ti-C_{sp}3-bond in the four membered ring.

Scheme 4

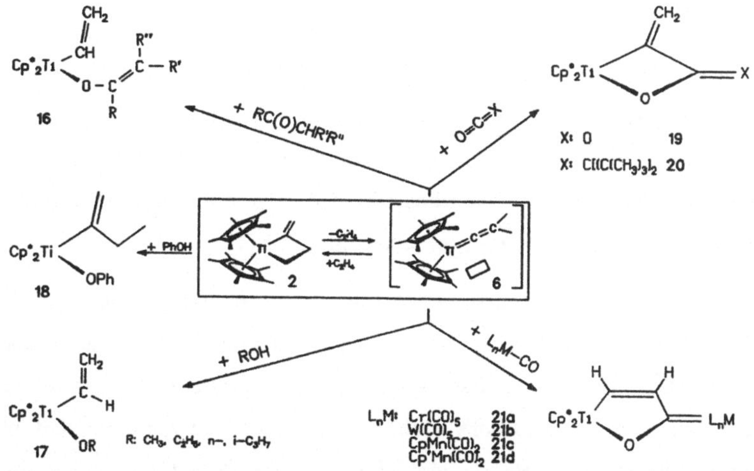

From this course of reaction it can be concluded, that the H-transformation and formation of 16, 17 is intermolecular, without participation of "tuck-in" species 14. This knowledge is valuable for the following cycloadditions.

So by using carbon dioxide or bis(tert-butyl)ketene as acceptor ligands, bordeaux-red (19) or golden-yellow crystals (20) are formed by cycloreversions, respectively. These complexes are characterized by NMR-spectroscopy and by mass spectra. The chemical shift of the exocyclic methylidene protons is of high diagnostic value of the four membered ring structure. In general, a singlett is observed for each proton [123].

Another type of NMR spectra has been obtained for the reaction products of the thermolysis of 2 in the presence of metal carbonyls as a heterocumulene-type molecules. So a typical AB-pattern is observed in the case of the reaction product of tungsten or chromium hexacarbonyl. From this behavior it can be concluded, that there is only a five-membered ring with two methine groups instead of one exocyclic methylidene group. That means that there must be a subsequent reaction or a rearrangement starting from the vinylidene intermediate. Consequently by using other metal carbonyls, such as cymantrene or methylcymantrene, similar carbene complexes (**21c**, **21d**) with an acetylene bridge between the titanium centre and the carbene C-atom are formed. These complexes are well crystallizing materials of high thermal stability (up to 220°C). In the ^{13}C-NMR spectra the typical signals of carbene C-atoms are observed (**21a**: 325.4, **21b**: 305.5, **21c**: 322.5, **21d**: 322.6 ppm). Additionally the signals of the carbonyl groups can be observed in the range of 230-240 ppm and of methine-C-atoms at 227 and 154 ppm (**21c**) with typical coupling values for the methine protons ($^1J_{CH}$: 156, 143 Hz) [123].

The structure of the titanaoxocyclic carbene complex with methylcymantrene was confirmed by single crystal X-ray diffraction (figure 6). So the nearly planar five membered titanaoxacyclic ring can be observed. Additionally, the titanium centre and the manganese atom are pseudotetrahedrally coordinated [123]. The Ti-O- and the Mn=C-bond are in the normal range of cyclic FISCHER-carbene complexes [124].

How can we explain this course of reaction, the formation of binuclear titanium-transition metal carbene complexes bridged by an acetylene group? That means that, a 1,2-H-shift must have taken place. In a first step the vinylidene intermediate is formed, followed by coordination of the metal-carbonyl stabilizing the coordinative unsaturated intermediate. In a further subsequent reaction the vinylidene is transformed into a coordinated acetylene (classical reaction of free vinylidenes [67, 68, 125, 126]) which undergoes cycloaddition with the metal carbonyl. This reaction is comparable with the reaction of zirconocene vinyl complexes with metal carbonyls forming five-membered cyclic FISCHER-carbene complexes [127].

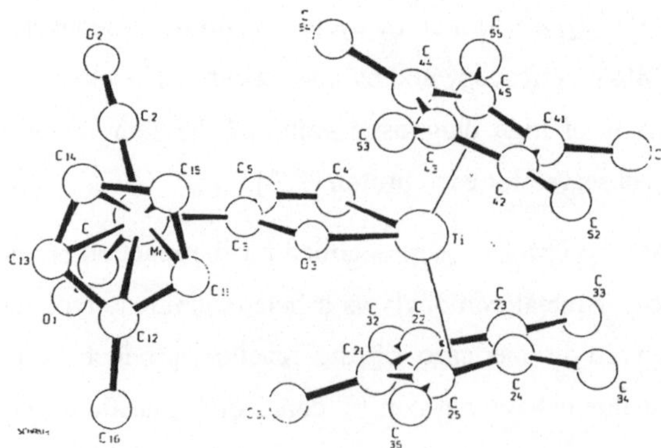

Figure 6. Perspective view of the molecular structure of **21d**

Acknowledgement

R. B. thanks the Alexander von Humboldt-Stiftung for a fellowship (1990/91, TU München) and Prof. Dr. P. Hofmann (TU München) for a fruitfull cooperation. Financial aid from the Fonds der Chemischen Industrie is gratefully acknowledged. Regarding structure analysis I am grateful to Dr. Trojanov (State University Moskau), to Dr. E. Herdtweck and Dr. P. Kiprof (TU München) and to Dr. U. Englert and B. Wagner (RWTH Aachen).

References

[1] N. Calderon, J. P. Lawrence, E. A. Ofstead, *Adv. Organomet. Chem.* **1979,** *17,* 449-492.

[2] R. H. Grubbs, *Progr. Inorg. Chem.* **1979,** *24,* 1-50.

[3] G. Erker, R. Nolte, R. Aul, S. Wilker, C. Krüger, R. Noe, *J. Am. Chem. Soc.* **1991,** *113,* 7594-7602.

[4] W. Röll, H.-H. Britzinger, B. Rieger, R. Zolk, *Angew. Chem.* **1990,** *102,* 339-341.

[5] J. J. W. Eshuis, Y. Y. Tan, A. Meetsma, J. H. Teuben, J. Renkema, G. G. Evens, *Organometallics* **1992,** *11,* 362-369.

[6] J. E. Bercaw, *J. Am. Chem. Soc.* **1974,** *96,* 5087-5095.

[7] J. M. Manriquez, D. R. McAlister, E. Rosenberg, A. M. Shiller, K. L. Williamson, S. I. Chan, J. E. Bercaw, *J. Am. Chem. Soc.* **1978,** *100,* 3078-3083.

[8] J. E. Bercaw, R. H. Marvich, L. G. Bell, H.-H. Brintzinger, *J. Am. Chem. Soc.* **1972,** *94,* 1219-1238.

[9] M. D. Fryzuk, T. S. Haddad, S. J. Rettig, *J. Am. Chem. Soc.* **1990,** *112,* 8125-8186.

[10] G. Erker, *Acc. Chem. Res.* **1984,** *17,* 103-109.

[11] P. T. Wolczanski, J. E. Bercaw, *Acc. Chem. Res.* **1980,** *13,* 121-127.

[12] J. Schwartz, J. A. Labinger, *Angew. Chem.* **1976,** *88,* 402-409.

[13] T. Gibson, *Organometallics* **1987,** *6,* 918-922.

[14] E. Negishi, *Acc. Chem. Res.* **1987,** *20,* 65-72.

[15] E. Negishi, D. E. van Horn, T. Yoshida, *J. Am. Chem. Soc.* **1985,** *107,* 6639-6647.

[16] M. L. H. Green, D. O'Hare, *Pure Appl. Chem.* **1985,** *57,* 1897-1910.

[17] M. Brookhart, M. L. H. Green, *J. Organomet. Chem.* **1983,** *250,* 395-408.

[18] M. E. Thompson, S. M. Baxter, A. R. Bulls, B. J. Burger, M. C. Nolan, B. D. Santarsiero, W. P. Schaefer, J. E. Bercaw, *J. Am. Chem. Soc.* **1987,** *109,* 203-219.

[19] H. Rabaa, J.-Y. Saillard, R. Hoffmann, *J. Am. Chem. Soc.* **1986,** *108,* 4327-4333.

[20] J. W. Lauher, R. Hoffmann, *J. Am. Chem. Soc.* **1976,** *98,* 1729-1742.

[21] P. J. Fagan, J. M. Manriquez, S. H. Vollmer, C. S. Day, V. W. Day, T. J. Marks, *J. Am. Chem. Soc.* **1981,** *103,* 2206-2220.

[22] L. Zhu, N. M. Kostic, *J. Organomet. Chem.* **1987,** *335,* 395-411.

[23] K. Tatsumi, A. Nakamura, P. Hofmann, P. Stauffert, R. Hoffmann, *J. Am. Chem. Soc.* **1985**, *107*, 4440-4451.

[24] G. Erker, K. Kropp, C. Krüger, A.-P. Chiang, *Chem. Ber.* **1982**, *115*, 2447-2460.

[25] G. Erker, F. Rosenfeldt, *J. Organomet. Chem.* **1982**, *224*, 29-42.

[26] G. Erker, R. Petrenz, *Organometallics* **1992**, *11*, 1646-1655.

[27] R. Beckhaus: *"Zur Chemie von Vinylverbindungen elektronenarmer Übergangsmetalle"* (Habilitation Theses), Fakultät für Naturwissenschaften Technische Hochschule Leuna-Merseburg **1989**.

[28] R. Beckhaus, K.-H. Thiele, *J. Organomet. Chem.* **1986**, *317*, 23-31.

[29] R. Beckhaus, K.-H. Thiele, *J. Organomet. Chem.* **1984**, *268*, C7-C8.

[30] R. Beckhaus, K.-H. Thiele, D. Ströhl, *J. Organomet. Chem.* **1989**, *369*, 43-54.

[31] R. Beckhaus, K.-H. Thiele, *Z. anorg. allg. Chem.* **1989**, *573*, 195-198.

[32] R. Beckhaus, S. Flatau, S. Trojanov, P. Hofmann, *Chem. Ber.* **1992**, *125*, 291-299.

[33] H. Yasuda, A. Nakamura, *Angew. Chem.* **1987**, *99*, 745-764.

[34] H. Yasuda, T. Okamoto, Y. Matsuoka, A. Nakamura, Y. Kai, N. Kanehisa, N. Kasai, *Organometallics* **1989**, *8*, 1139-1152.

[35] S. L. Buchwald, R. B. Nielsen, *Chem. Rev.* **1988**, *88*, 1047-1058.

[36] S. L. Buchwald, B. T. Watson, *J. Am. Chem. Soc.* **1987**, *109*, 2544-2546.

[37] S. L. Buchwald, R. T. Lum, J. C. Dewan, *J. Am. Chem. Soc.* **1986**, *108*, 7441-7442.

[38] S. L. Buchwald, R. T. Lum, R. A. Fisher, W. M. Davis, *J. Am. Chem. Soc.* **1989**, *111*, 9113-9114.

[39] J. B. Lee, G. J. Gajda, W. P. Schaefer, T. R. Howard, T. Ikariya, D. A. Straus, R. H. Grubbs, *J. Am. Chem. Soc.* **1981**, *103*, 7358-7361.

[40] D. A. Straus, R. H. Grubbs, *Organometallics* **1982**, *1*, 1658-1661.

[41] J. W. F. L. Seetz, B. J. J. van de Heisteeg, G. Schat, O. S. Akkerman, J. F. Bickelhaupt, *J. Mol. Catal.* **1985**, **28**, 71-83.

[42] R. D. Dennehy, R. J. Whitby, *J. Chem. Soc. Chem. Comm.* **1990**, 1060-1062.

[43] K. M. Doxsee, J. B. Farahi, H. Hope, *J. Am. Chem. Soc.* **1991**, *113*, 8889-8898.

[44] J. W. Park, L. M. Henling, W. P. Schaefer, R. H. Grubbs, *Organometallics* **1990**, *9*, 1650-1656.

[45] M. J. Burk, W. Tumas, M. D. Ward, D. R. Wheeler, *J. Am. Chem. Soc.* **1990**, *112*, 6133-6135.

[46] K. A. Brown-Wensley, S. L. Buchwald, L. Cannizzo, L. Clawson, S. Ho, D. Meinhardt, J. R. Stille, D. Straus, R. H. Grubbs, *Pure Appl. Chem.* **1983**, *55*, 1733-1744.

[47] S. L. Buchwald, R. H. Grubbs, *J. Am. Chem. Soc.* **1983**, *105*, 5490-5491.

[48] N. A. Petasis, E. I. Bzowej, *J. Am. Chem. Soc.* **1990**, *112*, 6392-6394.

[49] J. R. Stille, R. H. Grubbs, *J. Am. Chem. Soc.* **1983**, *105*, 1664-1665.

[50] S. H. Bertz, G. Dabbagh, C. P. Gibson, *Organometallics* **1988**, *7*, 563-565.

[51] L. Clawson, S. L. Buchwald, R. H. Grubbs, *Tetrahedron Letters* **1984**, *25*, 5733-5736.

[52] A. J. Lees, A. W. Adamson, *J. Am. Chem. Soc.* **1980**, *102*, 6876-6878.

[53] F. N. Tebbe, G. W. Parshall, D. W. Ovenall, *J. Am. Chem. Soc.* **1979**, *101*, 5074-5075.

[54] L. R. Gilliom, R. H. Grubbs, *J. Am. Chem. Soc.* **1986,** *108*, 733-742.

[55] T. M. Swager, R. H. Grubbs, *J. Am. Chem. Soc.* **1987,** *109*, 894-896.

[56] P. B. Mackenzie, K. C. Ott, R. H. Grubbs, *Pure Appl. Chem.* **1984,** *56*, 59-61.

[57] P. B. Mackenzie, R. J. Coots, R. H. Grubbs, *Organometallics* **1989,** *8*, 8-14.

[58] F. Ozawa, J. W. Park, P. B. Mackenzie, W. P. Schaefer, L. M. Henling, R. H. Grubbs, *J. Am. Chem. Soc.* **1989,** *111*, 1319-1327.

[59] E. V. Anslyn, R. H. Grubbs, *J. Am. Chem. Soc.* **1987**, *109*, 4880-4890.

[60] J. D. Meinhart, E. V. Anslyn, R. H. Grubbs, *Organometallics* **1989**, *8*, 583-589.

[61] J. B. Lee, K. C. Ott, R. H. Grubbs, *J. Am. Chem. Soc.* **1982,** *104*, 7491-7496.

[62] J. M. Hawkins, R. H. Grubbs, *J. Am. Chem. Soc.* **1988,** *110*, 2821-2823.

[63] W. C. Finch, E. V. Anslyn, R. H. Grubbs, *J. Am. Chem. Soc.* **1988**, *110*, 2406-2413.

[64] F. Bickelhaupt, *Angew. Chem.* **1987**, *99, 1020-1035.*

[65] H. J. R. de Boer, O. S. Akkerman, F. Bickelhaupt, G. Erker, P. Czisch, R. Mynott, J. L. Wallis, C. Krüger, *Angew. Chem.* **1986,** *98*, 641-643.

[66] G. Erker, P. Czisch, C. Krüger, J. M. Wallis, *Organometallics* **1985**, *4*, 2059-2060.

[67] P. J. Stang, *Chem. Rev.* **1978**, *78*, 383-405.

[68] M. M. Gallo, T. P. Hamilton, H. F. Schaefer III, *J. Am. Chem. Soc.* **1990**, *112*, 8714-8719.

[69] H. Werner, *Nachr. Chem. Tech. Lab.* **1992,** *40*, 435-444.

[70] M. I. Bruce, *Chem. Rev.* **1991,** *91*, 197-257.

[71] M. I. Bruce, A. G. Swincer, *Adv. Organomet. Chem. 1983, 22*, 59-128.

[72] R. D. Dennehy, R. J. Whitby, *J. Chem. Soc. Chem. Comm.* **1992,** 35-36.

[73] H. G. Alt, H. E. Engelhardt, M. D. Rausch, L. B. Kool, *J. Organomet. Chem.* **1987**, *329*, 61-67.

[74] M. M. Hills, J. E. Parameter, W. H. Weinberg, *J. Am. Chem. Soc.* **1987**, *109*, 597-599.

[75] E. S. Kline, Z. H. Kafafi, R. H. Hauge, J. L. Margrave, *J. Am. Chem. Soc.* **1987**, *109*, 2402-2409.

[76] E. L. Hoel, *Organometallics* **1986,** *5*, 587-588.

[77] V. C. Gibson, G. Parkin, J. E. Bercaw, *Organometallics* **1991**, *10*, 220-231.

[78] A. van Asselt, B. J. Burger, V. C. Gibson, J. E. Bercaw, *J. Am. Chem. Soc.* **1986**, *108*, 5347-5349.

[79] H. Werner, F. J. G. Alonso, H. Otto, J. Wolf, Z. Naturforsch. **1988,** 43b, 722-726.

[80] H. Werner, J. Wolf, G. Müller, C. Krüger, *Angew. Chem.* **1984**, *96*, 421-422.

[81] J. Wolf, H. Werner, O. Serhadli, M. L. Ziegler, *Angew. Chem.* **1983**, *95*, 428-429.

[82] C. Bianchini, M. Peruzzini, A. Vacca, F. Zanobini, *Organometallics* **1991**, *10*, 3697-3707.

[83] A. Davison, J. P. Selegue, *J. Am. Chem. Soc.* **1978**, *100*, 7763-7765.

[84] S. J. Landon, P. M. Shulman, G. L. Geoffroy, *J. Am. Chem. Soc.* **1985,** *107*, 6739-6740.

[85] A. Mayr, K. C. Schaefer, E. Y. Huang, *J. Am. Chem. Soc.* **1984**, *106*, 1517-1518.

[86] K. R. Birdwhistel, J. L. Templeton, *Organometallics* **1985,** *4*, 2062-2064.

[87] R. M. Bullock, *J. Am. Chem. Soc.* **1987**, *109*, 8087-8089.

[88] R. G. Beevor, M. J. Freeman, M. Green, C. E. Morton, A. G. Orpen, *J. Chem. Soc. Chem. Comm.* **1985**,

[89] R. G. Beevor, M. Green, A. G. Orpen, I. D. Williams, *J. Chem. Soc. Chem. Comm.* **1983,** 673-675.

[90] B. E. Boland-Lussier, M. R. Churchill, R. P. Hughes, A. L. Rheingold, *Organometallics* **1982,** *1*, 628-634.

[91] A. G. M. Barrett, N. E. Carpenter, M. Sabat, *J. Organomet. Chem.* **1988**, *352,* C8-C9.

[92] A. Wong, J. A. Gladysz, *J. Am. Chem. Soc.* **1982,** *104,* 4948-4950.

[93] D. R. Senn, A. Wong, A. T. Patton, M. Marsi, C. E. Strouse, J. A. Gladysz, *J. Am. Chem. Soc.* **1988,** *110,* 6096-6109.

[94] G. A. Luinstra, J. H. Teuben, *Organometallics* **1992**, *11,* 1793-1801.

[95] J. S. Merola, *Organometallics* **1989**, *8,* 2975-2977.

[96] A. Höhn, H. Otto, M. Dziallas, H. Werner, *J. Chem. Soc. Chem. Comm.* **1987**, 852-853.

[97] T. C. McKenzie, R. D. Sanner, J. E. Bercaw, *J. Organomet. Chem.* **1975,** *102,* 457-466.

[98] S. A. Cohen, P. R. Auburn, J. E. Bercaw, *J. Am. Chem. Soc.* **1983,** *105*, 1136-1143.

[99] R. J. McKinney, T. H. Tulip, D. L. Thorn, T. S. Coolbaugh, F. N. Tebbe, *J. Am. Chem. Soc.* **1981**, *103,* 5584-5586.

[100] R. R. Schrock, *Acc. Chem. Res.* **1979**, *12,* 98-104.

[101] K. H. Dötz, H. Fischer, P. Hofmann, F. R. Kreissl, U. Schubert, K. Weiss: *Transition Metal Carbene Complexes,* Verlag Chemie **1983**.

[102] R. Beckhaus, C. Zimmermann, B. Wagner, U. Englert, E. Herdtweck, *unpublished results* .

[103] W. R. Tikkanen, J. Z. Liu, J. W. Egan Jr., J. L. Petersen, *Organometallics* **1984**, *3,* 825-830.

[104] A. Kabi-Satpathy, C. S. Bajgur, K. P. Reddy, J. L. Petersen, *J. Organomet. Chem.* **1989**, *364,* 105-117.

[105] G. Erker, U. Korek, R. Petrenz, A. L. Rheingold, *J. Organomet. Chem.* **1991**, *421,* 215-231.

[106] W. E. Hunter, J. L. Atwood, G. Fachinetti, C. Floriani, *J. Organomet. Chem.* **1981**, *204,* 67-74.

[107] L. D. Durfee, I. P. Rothwell, *Chem. Rev.* **1988**, *88,* 1059-1079.

[109] G. Erker, R. Schlund, M. Albrecht, C. Sarter, *J. Organomet. Chem.* **1988**, *353,* C27-C29.

[110] G. Erker, R. Schlund, C. Krüger, *J. Organomet. Chem.* **1988,** *338,* C4-C6.

[111] M. Bochmann, L. M. Wilson, M. B. Hoursthouse, R. L. Short, *Organometallics* **1987**, *6,* 2556-2563.

[112] F. J. Berg, J. L. Petersen, *Organometallics* **1991**, *10,* 1599-1607.

[113] F. J. Berg, J. L. Petersen, Organometallics **1989,** 8, 2461-2470.

[114] K. Mashima, H. Haraguchi, A. Ohyoshi, N. Sakai, H. Takaya, *Organometallics* **1991**, *10,* 2731-2736.

[115] R. Beckhaus, I. Strauss, J. Runsink, *unpublished results* .

[116] C. McDade, J. C. Green, J. E. Bercaw, *Organometallics* **1982**, *1,* 1629-1634.

[117] R. Fandos, J. H. Teuben, G. Helgesson, S. Jagner, *Organometallics* **1991**, *10*, 1637-1639.

[118] L. L. Whinnery Jr., L. M. Henling, J. E. Bercaw, *J. Am. Chem. Soc.* **1991**, *113,* 7375-7582.

[119] M. J. Carney, P. J. Walsh, R. G. Bergman, *J. Am. Chem. Soc.* **1990,** *112*, 6426-6428.

[120] M. J. Carney, P. J. Walsh, F. J. Hollander, R. G. Bergman, *Organometallics* **1992**, *11,* 761-777.

[121] S. C. Ho, S. Hentges, R. H. Grubbs, *Organometallics* **1988**, *7,* 780-782.

[122] H. Werner, U. Brekau, *Z. Naturforsch.* **1989**, *44b,* 1438-1446.
68-70.

[123] R. Beckhaus, P. Kiprof, *unpublished results*.

[124] K. Mashima, K. Jyodoi, A. Ohyshi, H. Takaya, *J. Chem. Soc. Chem. Comm.* **1986**, 1145-1146.

[125] P. J. Stang, *Acc. Chem. Res.* **1982**, *15*, 348-354.

[126] R. Krishnan, M. J. Frisch, J. A. Pople, P. v. R. Schleyer, *Chem. Phys. Lett.* **1981**, *79*, 408-411.

[127] R. Beckhaus, K.-H. Thiele, *J. Organomet. Chem.* **1989**, *368*, 315-322.

Enantioselective Oligomerization of α-Olefins with Chiral Zirconocene/Aluminoxane Catalysts

Walter Kaminsky, Aurelia Ahlers, Oliver Rabe[1] and Wilfried König[2]

[1] Institute for Technical and Macromolecular Chemistry
[2] Institute for Organic Chemistry
University of Hamburg, D-2000 Hamburg 13, Germany

Summary

Chiral metallocenes can be used with methylaluminoxane to oligomerize olefins. The use of (S)-(1,1'-ethylenebis(4,5,6,7-tetrahydro-1-indenyl))zirconiumbis-(O-acetyl-(R)-mandelate) allowed the preparation of optically active branched olefins. The oligomerization can be controlled in such a way that mainly tri-, tetra-, and pentamers are produced. The optical rotation depends extremely on the oligomerization temperature and is nearly zero at 70 °C. At a reaction temperature of 20 °C trimeric propene (2,4-dimethyl-1-heptene) is obtained in the (S)-configuration with an optical purity of 97,5 %. The trimers could be separated analytically into pure enantiomers by gas-chromatography over a modified cyclodextrine phase. The tetramers are diastereomers and show higher optically rotation than the trimers.

1 Introduction

Highly active catalysts for the olefin polymerization and oligomerization are formed by combination of metallocenes and methylaluminoxane (MAO) [1-3]. Prochiral monomers like propene, 1-butene, 1-pentene, and 1-hexene could be polymerized to isotactic polymers with chiral metallocenes like (R,S)-1,1'-ethylene-bis-(4,5,6,7-tetrahydro-1-indenyl) zirconiumdichloride [4]. If, instead of the racemate, one of the two enantiomers of the transition metal compounding is used, optically active oligomers are obtained [5]. One way of producing oligomers is the hydrooligomerization [6-7]. In the presence of hydrogen the molecular weight of the products is low because of fast hydrogen transfer reactions controlling the chain length. The products are saturated hydrocarbons. Of greater interest are the more reactive optically active alkenes which could be synthesized by another way. The rate of chain termination relative to chain growth is increased by raising the catalyst concentration and simultaneously lowering the monomer concentration.

The separation of the racemic mixture of the chiral metallocene is possible by preparing diasteromers with optically active compounds like (R)(binaphthol) or O-actyl-(R)-mandelate. The mandelate was employed because its catalytic activity is similar to that of the corresponding dichloride and may be obtained in higher purity than the binaphtholate (Scheme 1).

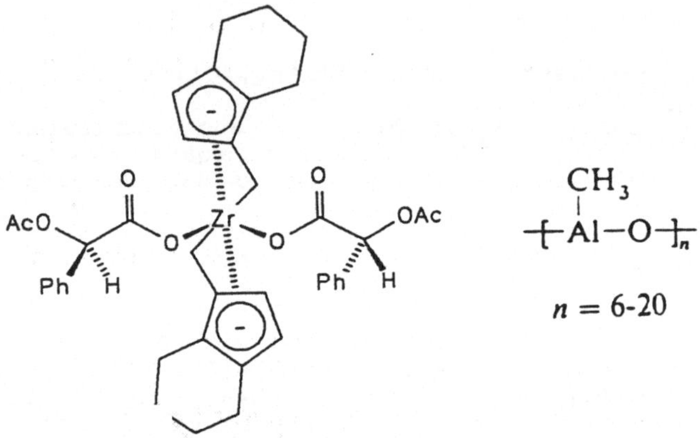

Scheme 1 Structural formula of (S)-[1,1'-ethylenebis(4,5,6,7-tetrahydro-1-indenyl)]zirconiumbis(O-acetyl-(R)-mandelate) and methylalumoxane

The oligomerization starts when an olefin undergoes insertion into a transition metal bond formed by methylation with methylalumoxane (Scheme 2).

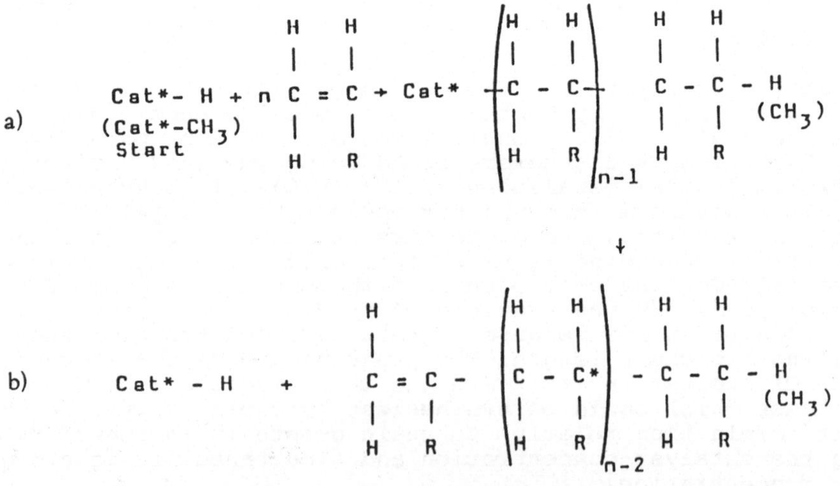

a) $\text{Cat*-H} \; (\text{Cat*-CH}_3) \; \text{Start} + n \, \text{CH}_2\text{=CHR} \rightarrow \text{Cat*} \left(\text{-CH}_2\text{-CHR-} \right)_{n-1} \text{CH}_2\text{-CHR-H} \; (\text{CH}_3)$

b) $\text{Cat* - H} + \text{CH}_2\text{=CR-} \left(\text{CH}_2\text{-C*HR} \right)_{n-2} \text{-CH}_2\text{-CHR-H} \; (\text{CH}_3)$

Scheme 2 Reaction Scheme for the Oligomerization of Olefins with Chiral Zirconocene Alumoxane Catalysts. a) Initiation and Propagation. b) Chain Termination

Subsequent insertions lead to chain growth. Chain termination takes place by β-hydrogen transfer to the transition metal atom or to a complex bound olefin, resulting in formation of the hydrid or alkyl transition metal compound in addition to the oligomer. The former, in turn, allows new insertion steps to occur. The formed dimers do not contain a chiral carbon atom. Optical activity is first observed in trimers and higher oligomers.

2 Propene Oligomerization

The average molecular weights of the produced oligopropenes can be controlled by adjustment of reaction temperature and monomer concentration. Ethylenebis(tetrahydroindenyl)zirconiumdi(O-acetyl-(R)-mandelate) was used as transition metal compound. To obtain sufficient amounts of product with a propene feed rate of 2,5 to 20 ml/min the reaction time had to be extended to between 15 and 24 hours. The reaction temperature was varied in the range from 20 to 60 °C (Table 1). A 1 l glass autoclave served as reaction vessel.

After a 15-minute-induction period during which metallocene and aluminoxane were allowed to prereact in 150 ml of thermostated toluene the propene gas was dosed into the stirred reaction mixture through a mass flow controller.

Table 1 Oligomerization of Propene using (S)-En($IndH_4$)$_2$Zr-($C_{10}H_8O_4$)$_2$/MAO
Zirconocene 5 x 10^{-4} mol/l in 150 ml Toluene
MAO 4,2 x 10^{-2} mol/l
Propene 10 ml/min by 2,2 bar Pressure

T (°C)	t (h)	Yield (g)	M_n (g/mol)	O_n
20	24	23	1590	38
30	24	22	524	13
40	24	24	403	11
50	19	18	251	6
60	21	21	219	

Depending on reaction times yields ranged from 18 to 24 g of oligopropenes. For reaction temperatures of 30 °C and above the products are oily liquids whereas oligomerization at 20 °C yielded waxy products. The reaction temperature has a great effect on the average degree of oligomerization O_n. Increasing the reaction temperature from 20 to 60 °C lowers the average molecular weight from 1540 to 219 g/mol. Under the given reaction conditions a variation of the monomer concentration did not have a significant effect on average molecular weights. In the concentration range from 2,5 to 20 ml/min of propene the average molecular weight rises from 233 to 371 g/mol.

The products contain oligopropenes of various degrees of oligomerization. By means of gas chromatography branched alkenes from dimers up to nonamers could be detected. Figure 1 shows the capillary gaschromatogram of a mixture of oligomers produced at 50 °C.

Table 2 Oligomerization of Propene of Different Monomer Concentrations. (S)-En$(IndH_4)_2$:5 x 10^{-4} mol/l; Methylaluminoxane 4,2 x 10^{-2} mol/l; T = 50 °C; t = 24 h

Propene [ml/min]	Yield [g]	M_{nJ} [g/mol]	O_n
2,5	5	233	6
5	12	211	5
10	23	236	6
15	35	333	8
20	46	371	9

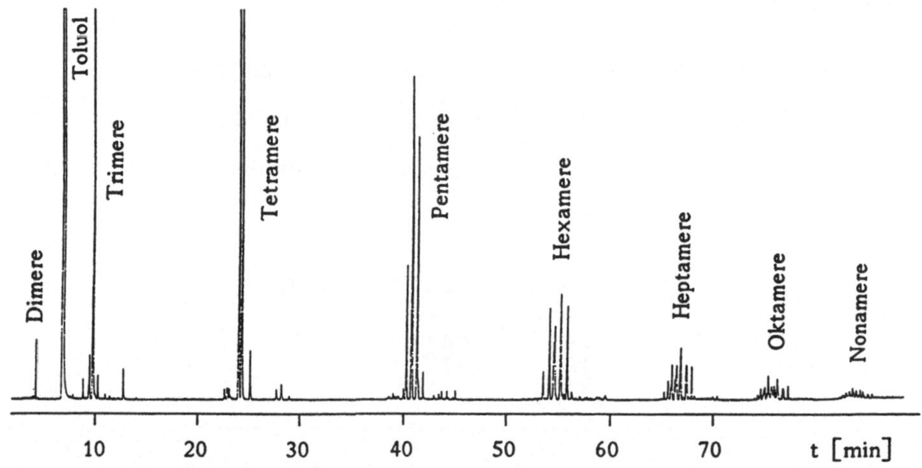

Figure 1 Gaschromatographic Separation of the Oligomerization Products of Propene Produced at 50 °C (50 m capillary column CP Sil 5 CB)

In the various oligopropene fractions a number of by-products (isomers) are formed next to the main component. Table 3 gives the amounts of olefins differing in degree of oligomerization. The table not only shows a shift of distribution maxima with temperature but also makes it clear that the oligomerization can be conducted in a way that mainly trimers through heptamers are formed. Provided that the number of active centers be independent of temperature the amounts of lower oligomers should grow with increasing temperature. Up to a reaction temperature of 60 °C this is the case. When the oligomerization is carried out at 70 °C, however, the sum of dimers to nonamers decreases indicating a slight decrease in the number of active centers.

Figure 2 shows the effect of the propene concentration on the mass distribution of the formed alkenes.

Table 3 Massdistribution of Propene Oligomers at Different Temperatures

T [°C)	Gew.-% Di	Tri	Tet	Pen	Hex	Hep	Oct	Non	Σ_{Di-Non}
30	0,1	2,0	3,1	4,5	4,3	4,1	4,2	4,1	26,4
40	1,3	5,2	6,9	8,3	8,4	7,3	6,9	5,2	49,5
50	2,4	8,9	11,5	12,2	11,3	9,4	5,8	2,8	64,3
60	4,0	14,4	18,8	17,2	12,7	9,4	6,1	2,9	85,5
70	3,5	11,5	17,1	16,2	14,5	11,5	6,0	2,5	82,8

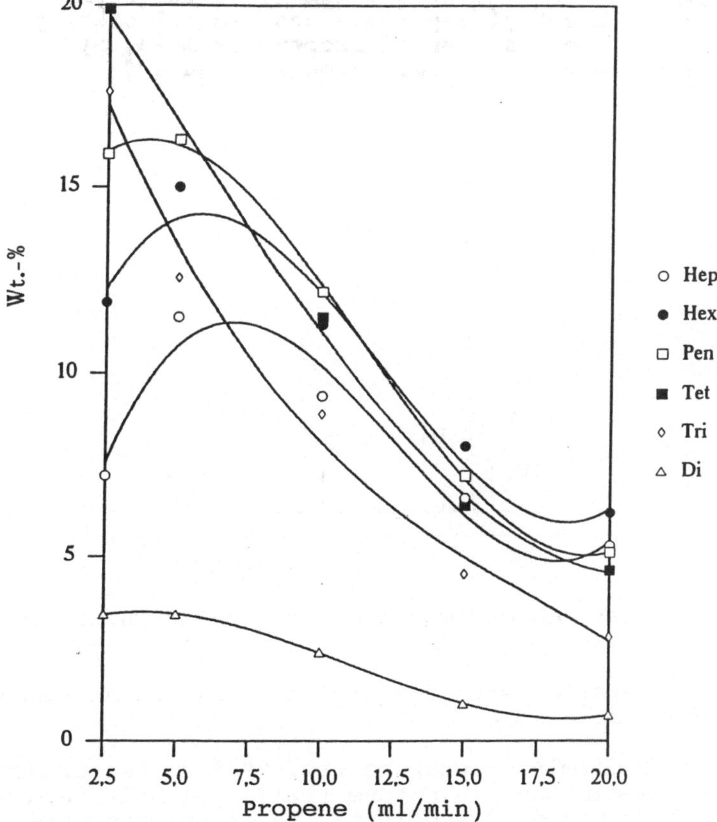

Figure 2 Weight-% of Different Oligomers versus the Propene Flow Rate

The contribution of the di-, tri- and tetramer fractions to the total alkene mixture produced in the reaction continuously increases with lower monomer concentrations. By contrast, the penta- through hexamers go through a maximum at about 5 ml/min of propene feed. Altogether, though, the portion of oligomers in the products show a marked increase with lower flowrates changing from 25 wt.-% for the sum of dimers to heptamers at 20 ml/min to 75 wt.-% at 2,5 ml/min.

3 Isomers

The propene oligomers synthesized with the (S)-En$(IndH_4)_2$Zr-$(C_{10}H_8O_4)_2$/MAO catalyst predominantly consist of 1-alkenes as they are formed by 1,2-insertion and isomeric by-products. In order to record all isomers of an individual degree of oligomerization the products were analytically separated by gas chromatography over a 50 m capillary column (cf. Fig. 1).

Independent of reaction temperature and monomer concentration the capillary gas chromatogram of the dimers features one main peak corresponding to 2-methyl-1-pentene and several other peaks of lower intensity. The fraction of propene trimers, by contrast, is made up of a mixture of several isomers (Figure 3).

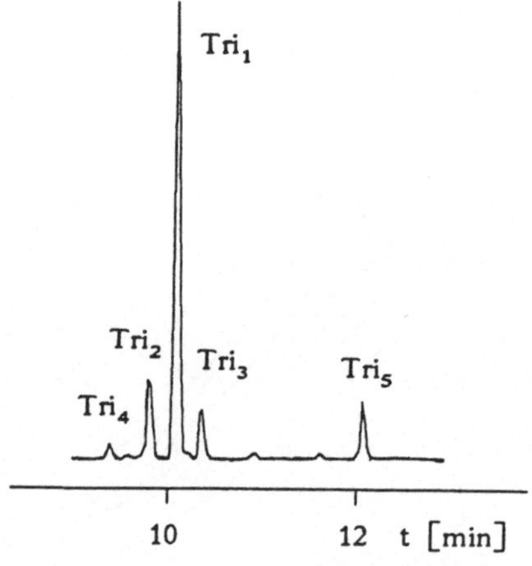

Figure 3 Capillary-gaschromatographic Separation of Propene Trimers synthesized by 50 °C

The composition varies with reaction temperature and monomer concentration (Table 4).

At 30 °C and 10 ml/min of propene feed 99 % of the trimer fraction consists of 2,4-dimethyl-1-heptene (Tri_1) which is formed by 1,2-insertions. A marked decrease in the relative concentration of this main component is observed at higher oligomerization temperatures as well as lower monomer feed rates. Simultaneously, other isomers that are formed by double bond migration, 2,1-, and 1,3-insertion gain significance. Scheme 3 assigns structures to the individual isomers.

Table 4 Distribution of Isomers in the Trimer Fraction of the Propene Oligomerization as a Function of Temperature and Propene Flow Rate
Catalyst: (S)-En$(IndH_4)_2Zr(C_{10}H_8O_4)_2$ 5 x 10^{-4} mol/l
MAO 4,2 x 10^{-2} mol al units/l
Reaction time: 24 h; propene pressure: 2,2 bar

T [°C]	Propene [ml/min]	Concentration wt.-%: Tri_1	Tri_2	Tri_3	Tri_4	Tri_5	ΣTri_{2-5}
30	10	99,3	-	-	-	0,7	0,7
40	10	96,5	1,3	0,4	-	0,9	2,8
50	2,5	85,4	5,9	5,9	1,1	2,0	14,9
50	5	87,3	5,5	3,3	1,1	2,1	12,0
50	10	88,2	4,7	1,9	0,9	2,2	9,7
50	15	89,2	4,7	1,3	-	2,2	8,2
50	20	91,8	4,3	1,2	-	2,2	8,7
60	10	80,2	7,3	4,3	1,5	5,2	18,3
70	10	71,1	10,2	6,9	2,3	6,5	25,9

Tri_1 2,4-Dimethyl-1-heptene

Tri_2 2,4-Dimethyl-2-heptene

Tri_3 2,6-Dimethyl-1-heptene

Tri_4 2,4,5-Trimethyl-1-hexene

Tri_5 2,4,6-Trimethyl-1-heptene

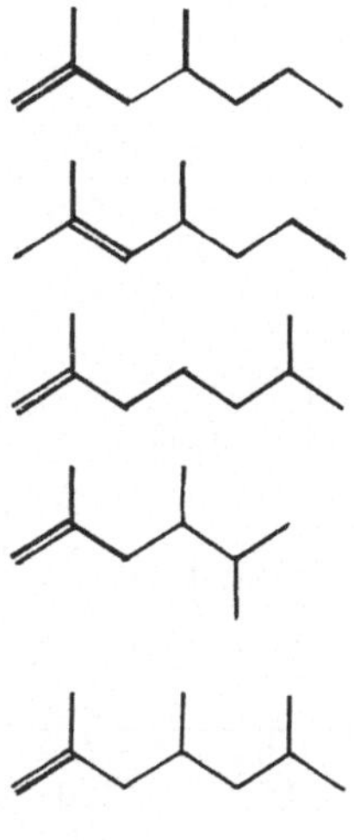

Scheme 3 Isomers of Trimeric Oligopropene

The trimer Tri_2 stems from Tri_1 via double bond migration. This reaction becomes increasingly important at higher temperatures.

Tri_3 is formed through an initial 2,1-insertion followed by a 1,3-insertion which, in turn, results from rearrangement of another 2,1-insertion after the first one.

This order of events becomes plausible when one considers that a regular 1,2-enchainment is sterically hindered after a 2,1-insertion thus favoring another 2,1-insertion. It is this steric hindrance between two adjacent methyl groups in a 2,1-1,2-sequence that is responsible for the relatively low concentration of Tri_4 which contains an initial 2,1-enchainment followed by two insertions with 1,2-orientation.

Finally, Tri_5 is formed as Tri_1 by three consecutive 1,2-insertions. This time, however, the initial propene unit is inserted into a Zr-methyl bond as it is formed in a reaction of the zirconocene with methylaluminoxane as opposed to a Zr-hydrogen bond resulting from the common β-hydride transfer. The isomers Tri_1, Tri_2, and Tri_5 were positively identified by NMR- and mass spectrometry. Figure 4 gives the ^{13}C-NMR spectrum of the main component (Tri_1).

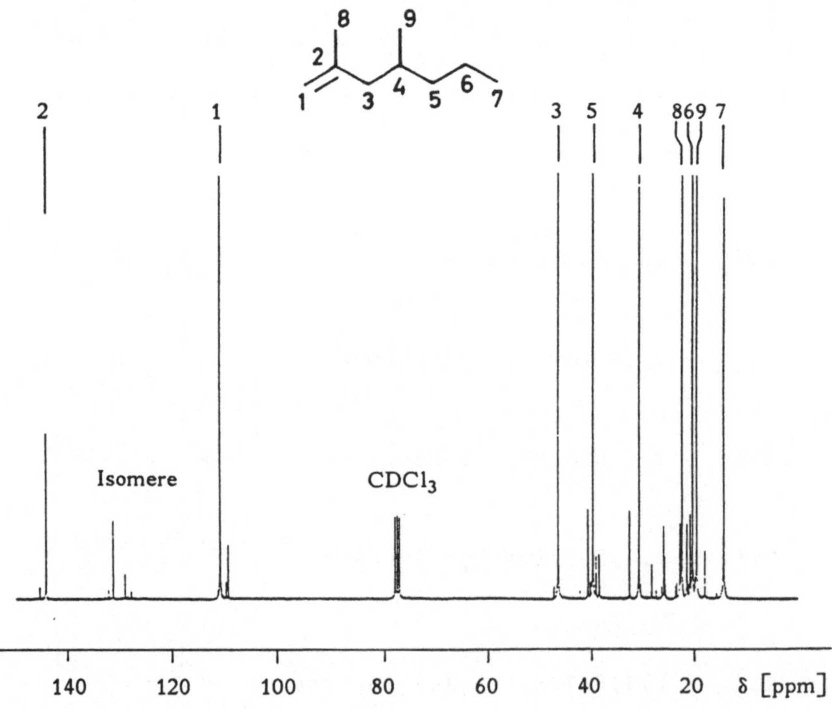

Figure 4 ^{13}C-NMR-Spectrum (90,6 MHz) of the Trimers of Oligopropene

A comparison of the NMR data with chemical shifts calculated after the increment methods of Grant and Paul [8] and Dorman [9] permitted assignment of the measured signals to structures. In no case do the calculated chemical shifts deviate more than 1 ppm from the measurements.

The propene tetramer contains two asymmetric carbon atoms. Therefore the synthesis with chiral metallocenes leads to the formation of diastereomers. They can be seen in the gas chromatogram of the tetramer fraction (cf. Fig. 1) as two principal peaks. As in the case of the trimers the main peaks of the tetramer fraction are accompanied by smaller ones stemming from double bond migration, 2,1-, and 1,3-insertions becoming more prominent at higher temperatures. Correspondingly, the composition of the pentamer and hexamer fractions of propene oligomers is growing even more complex. When the oligomerization is carried out at 30 °C or below, however, the diastereomers expected from exclusive 1,2-insertions account for more than 99 % of the products.

4 Optical Activity

The optical activity of the chiral oligopropenes was determined at various wavelengths. Polarimetric measurements were not only conducted with product mixtures from oligomerizations at various monomer concentrations and reaction temperatures but also with individual fractions of dimers, trimers, and tetramers. To this end the product mixtures were fractionated by distillation over a split tube column (Table 5).

Table 5 Specific Optical Rotation $[\alpha]^{25}$ of the Trimers, Tetramers and Mixed Oligomers at Different Wavelength and Different Reaction Temperatures and Propene Flow Rates

T (°C)	Propene ml/min	λ nm	Trimer $[\alpha]^{25}$	Tetramer $[\alpha]^{25}$	Oligomer $[\alpha]^{25}$
40	10	589	+ 1,7	+ 3,5	+ 3,0
40	10	546	+ 2,0	+ 4,3	+ 3,5
40	10	436	+ 3,7	+ 7,3	+ 5,8
40	10	365	+ 6,5	+11,8	+ 9,2
50	2,5	365			+ 4,9
50	5,0	365			+ 7,1
50	10	589	+ 0,9	+ 2,8	+ 2,6
50	10	365	+ 3,0	+ 8,7	+ 7,6
50	15	365			+ 5,6
50	20	365			+ 5,3
60	10	589	+ 0,4	+ 2,2	+ 2,2
60	10	365	+ 1,5	+ 6,6	+ 6,8
70	10	589	+ 0,08	+ 1,8	+ 1,9
70	10	365	+ 0,14	+ 5,6	+ 5,6

The propen oligomers starting with the trimers are dextro rotatory. As expected, the achiral dimer does not show any optical activity. The trimer, 2,4-dimethyl-1-heptene, which was produced with the (S)-En$(IndH_4)_2Zr(C_{10}H_8O_n)_2$/MAO catalyst, bears S-configuration since a specific optical rotation $[\alpha]_D^{24}$ of -6.1 was determined for the R-enantiomer [9].

With increasing reaction temperatures the specific optical rotation of all oligomers decays. This proves that the stereoselectivity of the organometallic catalyst decreases at higher temperatures. By contrast, the specific optical rotation increases with higher monomer feed rates going through a maximum at 10 ml propene/min. The optical activity of the tetramers is higher than that of the trimers. This increase is caused not only by the additional chiral carbon but also by an increase in stereoselectivity due to the longer alkyl chain attached to the active center. This difference is particularly significant at elevated temperatures. While the specific optical rotation of the trimer is lowered by a factor of 20 in the temperature intervall from 40 to 70 °C it is only reduced by one half for the tetramer.

The extent of stereoselectivity in the chiral synthesis can be checked by determining the enantiomeric excess of the optically active alkenes in the products. Since no literature data was available for the optical rotation of the enantiomerically pure alkenes their optical purity was determined through gaschromatographic resolution of enantiomers by means of an optically active column.

Thermostable substituted cyclodextrines are best suited as asymmetric phases [10, 11]. The trimer, 2,4-dimethyl-1-heptene, was resolved into its enantiomers by capillary gaschromatography with an octakis-(6-O-methyl-2,3-d-O-pentyl)-γ-cyclodextrine phase. In addition to that the racemate of a trimer fraction synthesized with the racemic metallocene rac-En$(IndH_4)_2(_2ZrCl_2)$ was separated by the same asymmetric GC-column.

At low temperatures (20 °C) the formation of the first chiral center proceeds with a high selectivity of 97,6 % leading to an enantiomeric excess of 95,3 %. At higher temperatures the ee-value decreases to 23,8 % at 50 °C and 2,5 % at 70 °C. As expected, the ee-value of the trimer produced with the racemic catalyst is 0 (Figure 5).

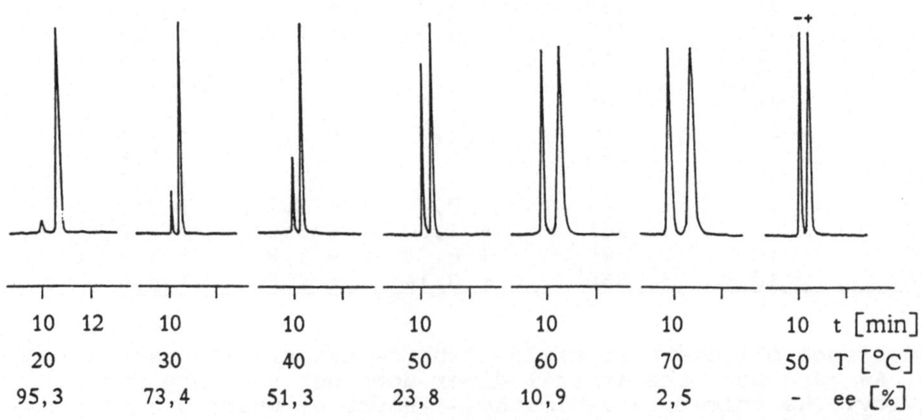

Figure 5 Asymmetric Oligomerization of Propene. Gaschromatographic Separation of 2,4-Dimethyl-2-heptene(Trimer) using Octakis(6-o-methyl-2,3-di-O-pentyl)cyclodextrine

5 1-butene and 1-pentene Oligomerization

The oligomers of 1-butene and 1-pentene can be synthesized at similar temperatures and monomer concentrations as the propene compounds. Normally it is easier to obtain low molecular weights with longer chained 1-olefins. Table 6 shows the specific optical rotation for oligomers of 1-butene. As catalyst (S)-En($IndH_4$)$_2$-$ZrCl_2$/MAO as it is obtained by splitting of the binaphthol complex with hydrogen chloride is used.

Table 6 Specific Optical Rotation at Different Wave Lengths of 1-butene Oligomers Prepared using (S)-(+)-En($IndH_4$)$_2$-$ZrCl_2$/MAO at 50 or 70 °C

Wave Length:	Φ 589 nm	Φ 365 nm
Dimer	0	0
Trimers (50 °C)	- 1,0	- 3,5
Trimers (70 °C)	- 0,3	- 0,8
Tetramers (50 °C)	- 3,2	- 10,0
Tetramers (70 °C)	- 1,2	- 3,4

As predicted, there is no optical rotation in the dimer fraction. The optical rotation of the higher oligomers decreases with increasing oligomerization temperature. The optical rotation of the tetramers is again higher than that of the trimers. Contrary to the propene oligomers the butene oligomers are levorotatory. The optical purity (ee) of the oligomers as a function of reaction temperature is shown in Figure 6.

The ee-value increases from 14 % at high oligomerization temperatures of 60 °C up to 92 % at low temperatures (20 °C).

Also the next homologue, 1 pentene, could be oligomerized with the asymmetric ethylene-bridged (S)-Et-($IndH_4$)$_2$$ZrMe_2$/MAO-system. The product mixtures as well as fractions of defined degree of oligomerization show measurable optical activity. Specific optical rotation values are of the same order of magnitude as in the case of 1-butene oligomers. Table 7 lists molar optical rotation values for pentene oligomers synthesized with the optically active (S)-Et($IndH_4$)$_2$$ZrMe_2$ catalyst.

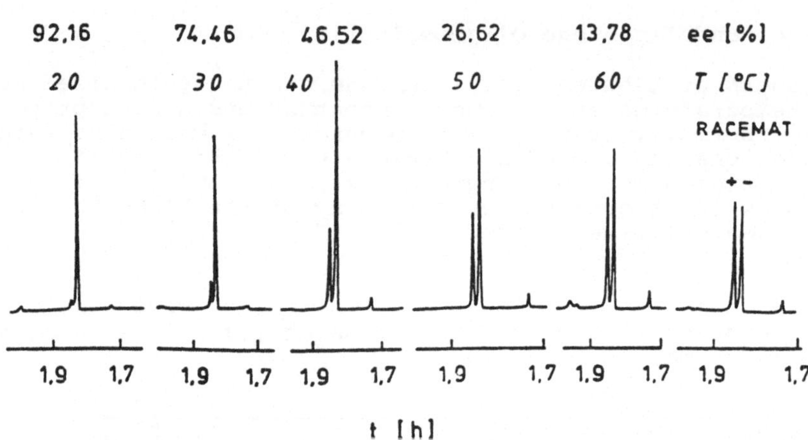

Figure 6 Gaschromatographic Separation of 2,4-Diethyl-1-octen (Trimer of butene) with Octakis(6-O-methyl-2,3-di-O-pentyl)cyclodextrine. Glass Capillary Column = 50 m, T = 55 °C, Gas = 1 bar H_2

Table 7 Molar Rotation of Pentene Oligomers. Catalyst: (S)-(+)-Et$(IndH_4)_2ZrCl_2$/MAO at 10 °C

Component	Wave Length:	589 nm	546 nm	365 nm
Dimers		-	-	-
Trimers		- 1,33	- 1,65	- 5,07
Tetramers		- 2,59	- 3,41	- 8,11
Product Mixture (M_n = 290 g/mol)		- 1,86	- 2,15	- 5,58

As expected, the optical rotation first increases with increasing number of chiral centers in the molecule and then declines again after going through a maximum. This behavior is due to the lower degree of asymmetry of the chiral carbon atoms in longer chained molecules.

The product molecular weights were confirmed by GC/MS and is in accord with the expected highly isospecific 1,2-insertion and termination by β-hydride transfer (Scheme 4).

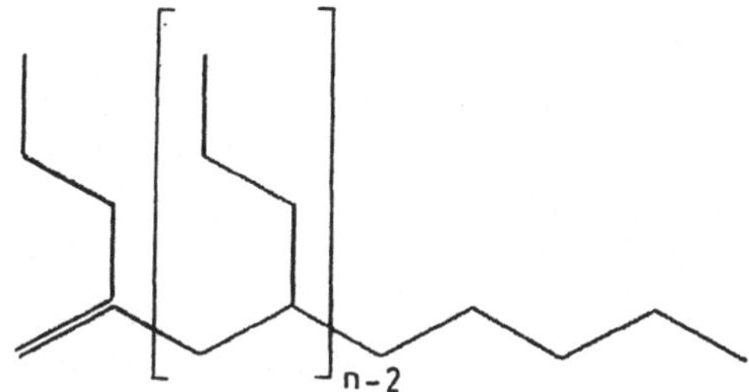

Scheme 4 Structure of 1-pentene Oligomers; n: degree of oligomerization

5 Conclusion

The enantioselective oligomerization of 1-olefins is possible with chiral metallocene/MAO catalysts. Only traces (10^{-4} - 10^{-6} mol/l) of pure enantiomers or diasteromers of the organometallic compounds are necessary to obtain moles of optically active, branched alkenes with high ee-values. Under identical reaction conditions the tetramers show higher optical rotation than the trimers. This shows that for a high stereoselectivity in Ziegler-Natta catalyses chirality about the central atom of the active site is mandatory. The growing polymer chain, however, adds significantly to the overall stereospecifity of the insertion step, thus raising the isotacticity of the polymeric product.

6 References

[1] H. Sinn, W. Kaminsky, Adv. Organomet.Chem. 18 (1980) 99

[2] W. Kaminsky, in R.P. Quirk (ed.): Transition Metal Catalyzed Polymerization, harwood academic publishers, New York 1983, p 225

[3] W. Kaminsky, H. Lücker, Makromol.Chem.Rapid Commun. 5 (1984) 225

[4] W. Kaminsky, K. Külper, H.H. Brintzinger, F.R.W.P. Wild, Ang. Chem. 97 (1985) 507; Angew.Chem.Int.Ed.Engl. 24 (1985) 507

[5] P. Pino, P. Cioni, J. Wei, J.Am.Chem.Soc. 109 (1987) 6189

[6] P. Pino, P. Cioni, M. Galimberti, J. Wei, N. Piccolrovazzi, in W. Kaminsky, H. Sinn (eds.): Transition Metals and Organometallics as Catalysts for Olefin Polymerization, Springer, Berlin 1988, p. 269

[7] W. Kaminsky, A. Ahlers, N. Möller-Lindenhof, Angew.Chem. 101 (1989) 1304; Angew.Chem.Int.Ed.Engl. 28 (1989) 1216

[8] D.M. Grant, E.G. Paul, J.Am.Chem.Soc. 86 (1964) 2984

[9] D.E. Dorman, M. Jantelat, J.D. Roberts, J.Organomet.Chem. 36 (1971) 2757

[10] P.A. Levene, R.E. Marker, J.Biol.Chem. 92 (1931) 455

[11] W.A. König, Nachr.Chem.Tech.Lab. 37 (1989) 471

[12] W.A. König, S. Lutz, G. Wenz, Angew.Chem. 100 (1988) 989; Angew.Chem.Int.Ed.Engl. 27 (1988) 979

Recent Advances in Organopalladium Chemistry

Jan-E. Bäckvall

Department of Organic Chemistry, University of Uppsala
Box 531, S-751 21 Uppsala, Sweden

Summary

Intramolecular palladium-catalyzed 1,4-oxidation of conjugated dienes leads to annulation type reactions and spirocyclizations. The chiral starting materials for the annulation type reactions were prepared in enatiomerically pure form. Enzymatic hydrolysis of *meso*-1,4-diacetoxy-2-cycloalkenes and subsequent palladium-catalyzed reactions led to *both enantiomers* of (cyclohexa-2,5-dienyl)acetic acid and (cyclohepta-2,5-dienyl)acetic acid. Finally, a palladium-catalyzed tandem cyclization of ω-amido-1,3-dienes resulting in pyrrolizidine and indolizidine structures was developed.

Regio- and stereoselective functionalization of conjugated dienes is a challenging problem in organic chemistry. Apart from the Diels-Alder reaction [1], which involves the 1,4 addition of an alkene to a conjugated diene, very few such functionalizations are known. Related reactions such as addition of singlet oxygen [2] and nitroso compounds [3] to conjugated dienes also lead to selective 1,4-syn addition.

Electrophilic additions to conjugated dienes normally give poor regio- and stereoselectivity. For example, halogenation produces mixtures of 1,2- and 1,4-addition products[4,5] and if the 1- and 4-positions are prochiral mixtures of stereoisomers are obtained [5]. Recently, we developed a number of palladium(II)-catalyzed oxidations of conjugated dienes in which two nucleophiles are added across the diene (eq. 1) [6-8]. In these reactions carboxylates, halides or alkoxides are added to the diene in the presence of an oxidant and a palladium(II) catalyst. These reactions, which proceed via (π-allyl)palladium

$$\text{diene} + X^- + Y^- \xrightarrow[\text{oxidation}]{\text{cat. Pd(II)}} X\text{–CH}_2\text{CH=CHCH}_2\text{–}Y \quad (1)$$

X = OAc, OOCR, OR
Y = Cl, OAc, OOCR, OR

intermediates, are highly regioselective and in systems where both the 1- and 4-positions are prochiral, the additions are stereoselective. The reactions have been applied to the synthesis of natural products in both cyclic and acyclic systems [9-11].

Recently the palladium-catalyzed 1,4-oxidations of cyclic 1,3-dienes were extended to intramolecular versions (Scheme 1) [12-14]. Two types of intramolecular 1,4-oxidations were developed; an annulation type reaction and a spirocyclization. In the former type of reaction

Scheme 1

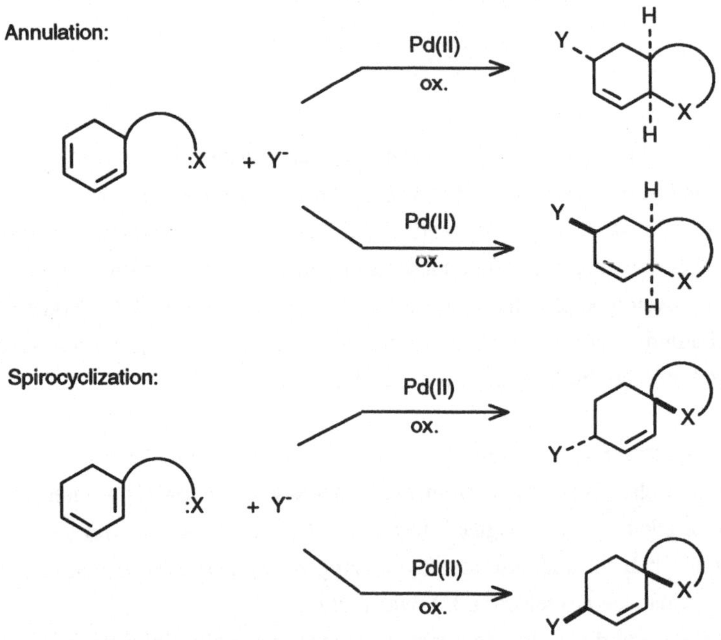

amides, carboxylic acids and alcohols were used as nucleophiles. Thus, fused pyrrolidines, lactones tetrahydrofurans, and tetrahydropyrans were obtained with dual control of the stereochemistry in the addition step. In the spirocyclization alcohols served as nucleophile in the intramolecular attack, resulting in stereoselective oxaspirocyclization. One example of the high regio- and stereoselectivity and also product selectivity in the annulation type reaction is given in Scheme 2.

The intramolecular version of the palladium-catalyzed 1,4-oxidation also allows the use of nitrogen nucleophiles. Thus, fused [7,5] and [6,5] pyrrolidines were obtained in regio- and stereoselective reactions [13]. The formation of a hexahydroindole (**1**) with defined stereochemistry was recently applied to the synthesis of lycorane alkaloids (Scheme 3) [15]. By switching the order of hydrogenation and Bischler-Napieralsky cyclization in the transformation of **2** to **3**, the α- or γ-lycoranes were obtained with complete stereoselectivity (>99%). Bischler-Napieralsky cyclization of **2** resulted in a highly stereoselective epimerization.

In the palladium-catalyzed intramolecular 1,4-oxidation of conjugated dienes products containing three chiral centra are obtained with full control of stereochemistry. It would

Scheme 2. Pd(II)-catalyzed oxidation to fused tetrahydrofurans

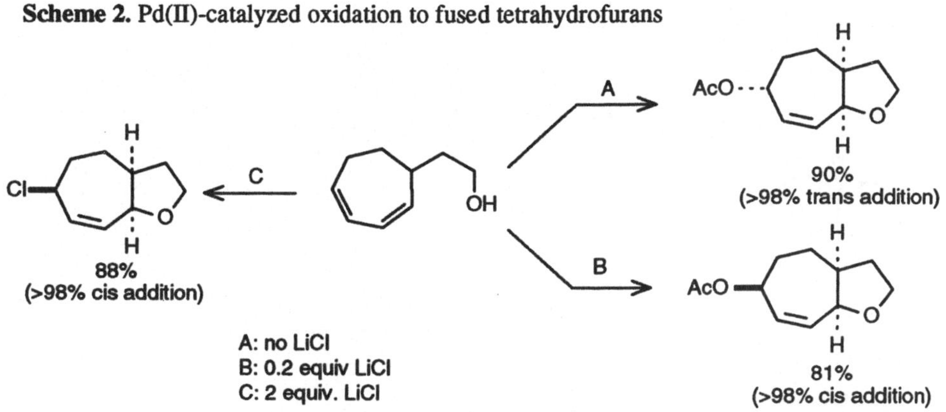

Scheme 3

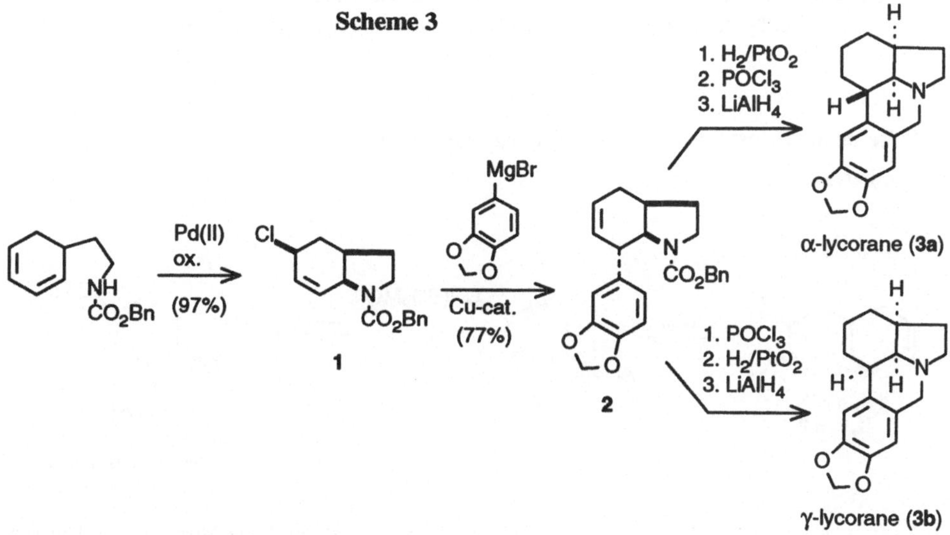

therefore be of great synthetic interest to use an enantiomerically pure 5-substituted 1,3-diene as the starting material (Scheme 4). In this way a number of highly functionalized enantiomerically pure heterocyclic compounds would be available.

(Cycloalka-2,4-dienyl)acetic acids (**4**) are one of the starting materials for the intramolecular reactions depicted in Scheme 1. They are also the precursors for the other 5-substituted 1,3-dienes used as starting materials in the intramolecular palladium-catalyzed 1,4-oxidations. A retrosynthetic analysis for the preparation of chiral non-racemic diene acids **4** is given in Scheme 5. The acid **4** can be obtained from allylic acetate **5** via elimination of acetic acid and decarboxylation. Compound **5** could in turn be obtained from meso diacetate **6** via an enantioselective reaction. Such an enantioselective reaction could in principle utilize a chiral palladium(0) catalyst and dimethyl malonate as the nucleophile. Enantioselective

Scheme 4. Annulation type reaction

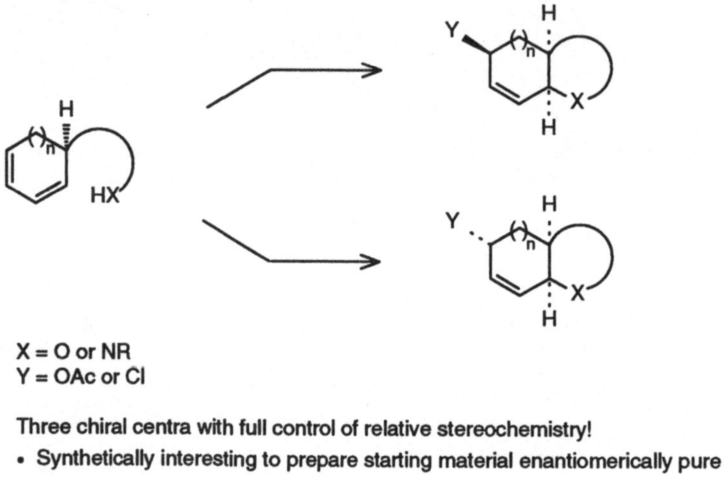
H
HX
Y
H
X
H
Y
H
X
H
X = O or NR
Y = OAc or Cl
Three chiral centra with full control of relative stereochemistry!
• Synthetically interesting to prepare starting material enantiomerically pure

Scheme 5. Retrosynthetic analysis for enantioselective synthesis of chiral diene acids

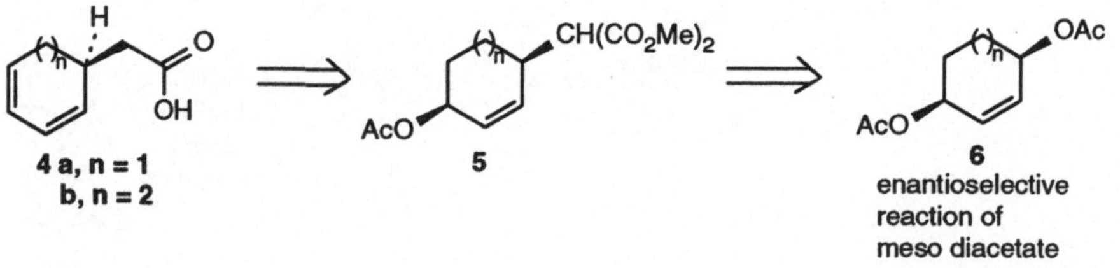
H
O
OH
4 a, n = 1
b, n = 2
CH(CO2Me)2
AcO
5
OAc
AcO
6
enantioselective
reaction of
meso diacetate

Pd(0)-catalyzed nucleophilic substitution on related meso systems have been reported in the literature [16]. However, the method chosen by us for the transformation of **6** to **5** involves an enzymatic hydrolysis of the meso diacetate to the corresponding *cis*-4-acetoxy-2-cycloalkenol.

The diacetate **6a** required for the enzymatic hydrolysis was obtained through a Pd-catalyzed diacetoxylation [6]. Initial attempts to obtain enantioselective enzymatic hydrolysis of **6a** were discouraging. For example the use of acetylcholine esterase (ACE) gave only low optical yields (15% ee) at about 40 % conversion. In substrate **6a** the -CH=CH- side and CH_2-CH_2 side are probably too similar such that the enzyme cannot distinguish between the two enantiomeric approaches [17]. In order to obtain high enantioselectivity in the enzymatic hydrolysis it is therefore necessary to brominate the double bond [17], which makes one side very different from the other. After hydrolysis the

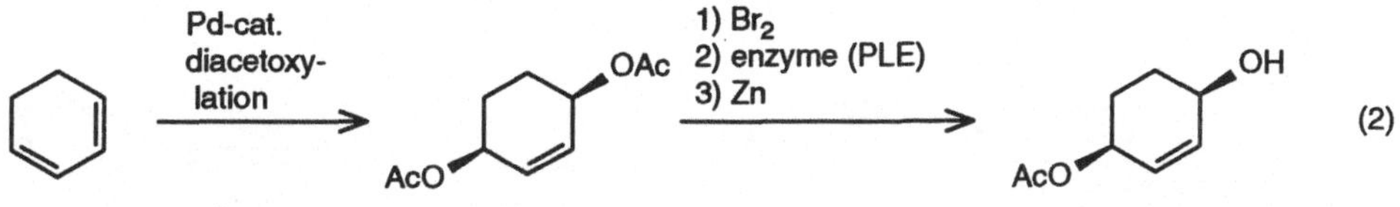

Pd-cat. diacetoxylation → 6a; 1) Br_2 2) enzyme (PLE) 3) Zn → 7a (> 98% ee) (2)

bromine is removed by zinc (eq. 2), and in this way monoacetate **7a** is obtained in >99% ee.

The chiral non-racemic monoacetate **7a** is a useful synthon for further functionalization, in particular via Pd(0)-catalyzed reactions [18]. Thus, a palladium(0)-catalyzed nucleophilic substitution of the allylic acetate in **7a** and subsequent acetylation leads to the R-acetate (Scheme 6). If on the other hand the carbonate is made from the alcohol followed by a palladium-catalyzed nucleophilic substitution the S-acetate is produced. It is known that an allylic carbonate is considerably more reactive than an allylic acetate, resulting in a complete product selectivity in the palladium-catalyzed reaction. In this way both enantiomers are available from the same intermediate.

Scheme 6. Compound **7a** as a useful synthon

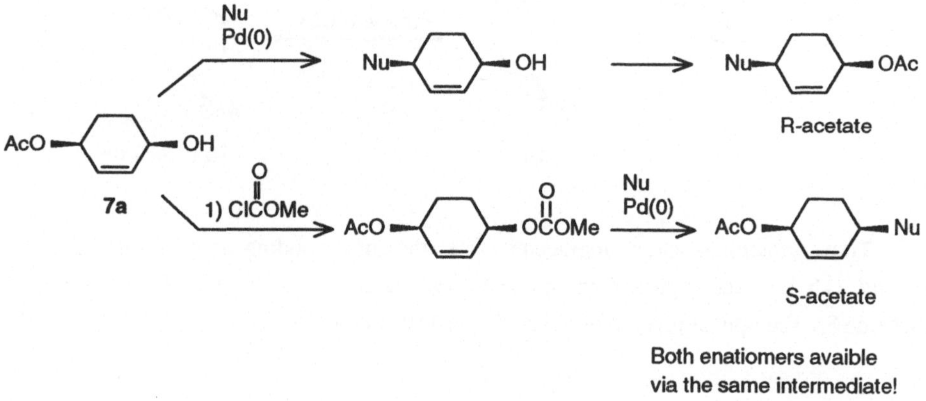

Nu, Pd(0); 1) ClCOMe; Nu, Pd(0); R-acetate; S-acetate

Both enatiomers avaible
via the same intermediate!

This methodology was applied to the synthesis of both enantiomers of (cyclohexa-2,4-dienyl)acetic acid (Scheme 7) [19]. Transformation of the alcohol group of **7a** to a carbonate followed by a palladium-catalyzed reaction with sodium dimethyl malonate gave **8a**. Subsequent Pd(0)-catalyzed elimination of acetic acid and decarboxylation afforded diene acid **9a**, which had an optical rotation of +196°. A direct Pd(0)-catalyzed malonate substitution of acetate in **7a** followed by acetylation afforded **10a**. Transformation of **10a** in the same manner as described for the enantiomer **8a** afforded diene acid **11a** which had an optical rotation of -197°.

The two enantiomers of the analogous (cyclohepta-2,4-dienyl)acetic acid were prepared

Scheme 7

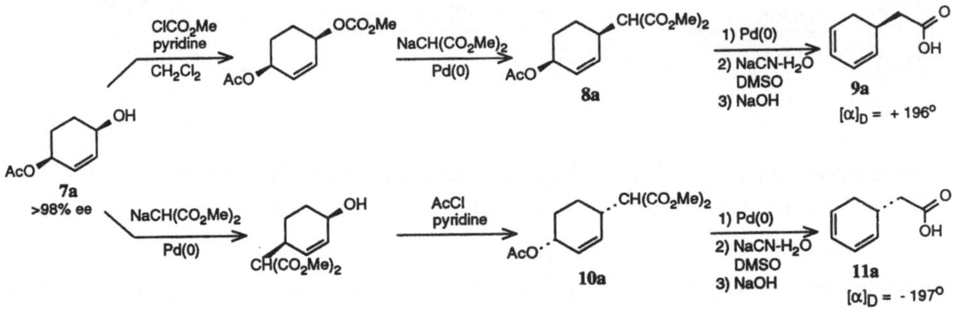

by employing the same methodlogy. Palladium-catalyzed diacetoxylation [6] of 1,3-cycloheptadiene afforded the *cis*-diacetate **6b**, which underwent enzymatic hydrolysis catalyzed by acetylcholine esterase [20]. In this case the two sides (-$CH_2CH_2CH_2$- and CH=CH) are sufficiently different to allow the direct enzymatic hydrolysis without prior functionalization of the double bond. The monoacetate obtained was > 98 % ee (eq. 3).

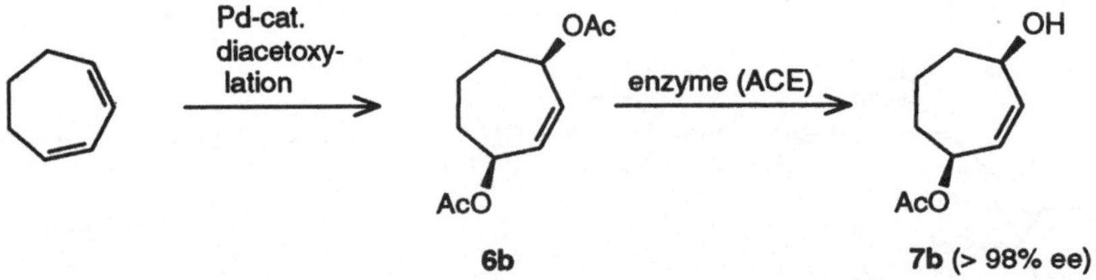

(3)

Transformation of chiral nonracemic **7b** to the corresponding enantiomeric diene acids **9b** and **11b** was accomplished in the same way as described for the six-membered ring (Scheme 8). The optical rotation of acids **9b** and **11b** was + 192° and -197°, respectively.

Scheme 8

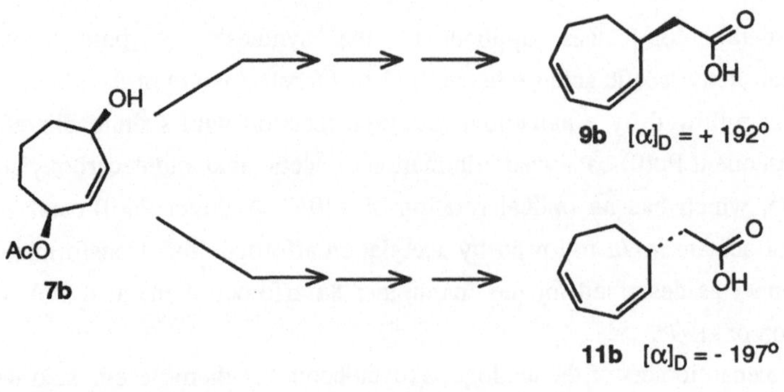

The enantiomers **9a** and **11a** of (cyclohexa-2,4-dienyl)acetic acid were cyclized in a palladium-catalyzed *trans*-acetoxy-lactonization reacion two give the two enantiomeric lactones **12a** and **13a** respectively (eqs 4 and 5). Analyses of the lactones by hydrolysis of the

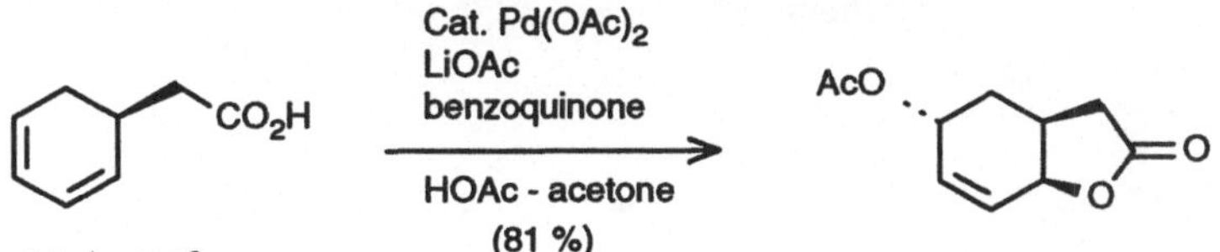

Cat. $Pd(OAc)_2$, LiOAc, benzoquinone; HOAc - acetone (81 %) (4)

$[\alpha]_D$=+196° — **12a** R-acetate, 93% ee

as obove (5)

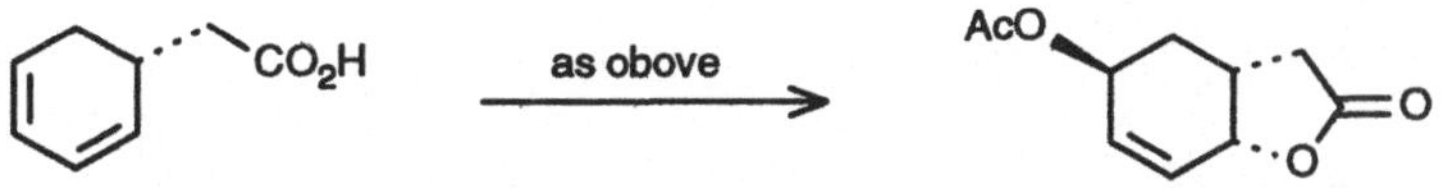

$[\alpha]_D$=-197° — **13a** S-acetate, 93% ee

acetate and subsequent transformation into Mosher's esters showed that the optical purity was 93 % ee in both cases.

A further extension of the palladium-catalyzed intramolecular 1,4-oxidations would be to utilize a nucleophile with the ability of making a two-fold nucleophilic attack. This would lead to a synthetically useful transformation and constitute a formal [4+1] cycloaddition (Scheme 9). If the nucleophile is nitrogen such a reaction would create 1-azabicyclic systems

Scheme 9. Formal [4 + 1] Cycloaddition

Nu — Pd(II), oxidation → Nu

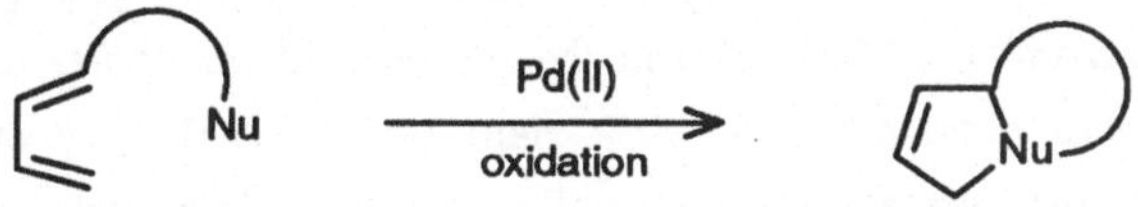

which have the structural units of pyrrolizidine and indolizidine alkaloids [21]. Some examples of such alkaloids are given in Figure 1.

Figure 1. Pyrrolizidine and Indolizidine Alkaloids

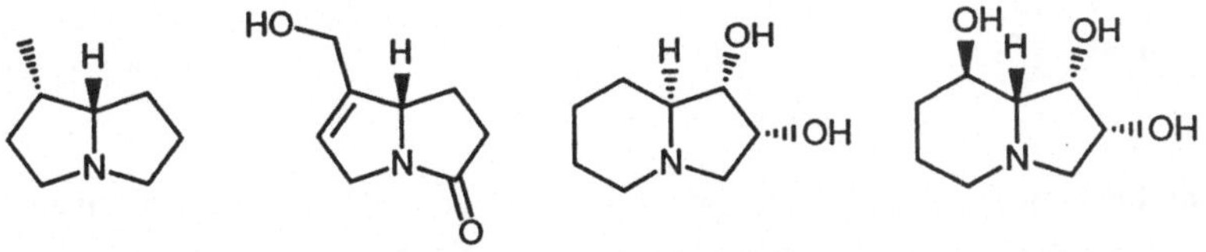

Heliotridane *Supinidine* *Indolizidinediol* *Swainsonine*

The starting materials for the planned intramolecular tandem cyclization were prepared as shown in Scheme 10. The readily available diene acid **14** was transformed into amide **15** via an CN^--catalyzed reaction with ammonia according to Högberg [22]. The chain

Scheme 10

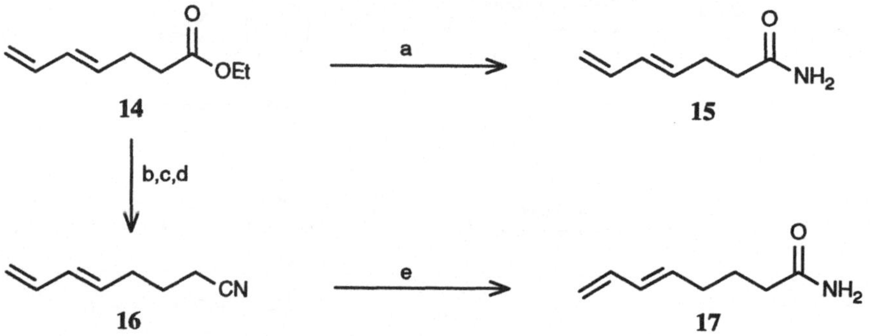

(a) NH_3, NaCN, MeOH, 85%; (b) DIBALH, CH_2Cl_2, 95%; (c) MsCl, Et_3N, THF, 96%; (d) NaCN, THF / DMSO, 95%; (e) KOH, t-BuOH, 67%.

elongation of **14** with one carbon was carried out using standard procedures to give nitrile **16**, which on selective hydrolysis afforded amide **17**. Attempted Pd-catalyzed intramolecular reactions of amides **15** and **17** with oxidants such as *p*-benzoquinone, urea hydroperoxide or isoamyl nitrite gave poor yields of cyclized product. After examining variations of solvents and oxidants we found that cupric chloride in tetrahydrofuran (THF) gave good results. When this reaction was performed at 60°C and in the presence of molecular oxygen the desired bicyclic products were obtained in 85-90% yield (eq. 6) [23]. The mechanism of this cyclization involves the formation of a (π-allyl)palladium intermediate via nucleophilic

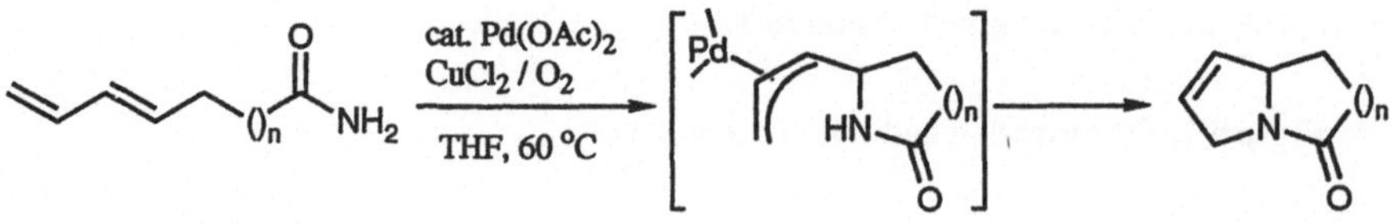

(6)

attack by the amide nitrogen on a (diene)palladium complex. Subsequent nucleophilic attack by nitrogen on the π-allyl group would produce the bicyclic compound.

The palladium-catalyzed tandem cyclization was applied to the synthesis of (±)-heliotridane (Scheme 11) [23]. Diene **18** was transformed into the amide and subsequent

Scheme 11

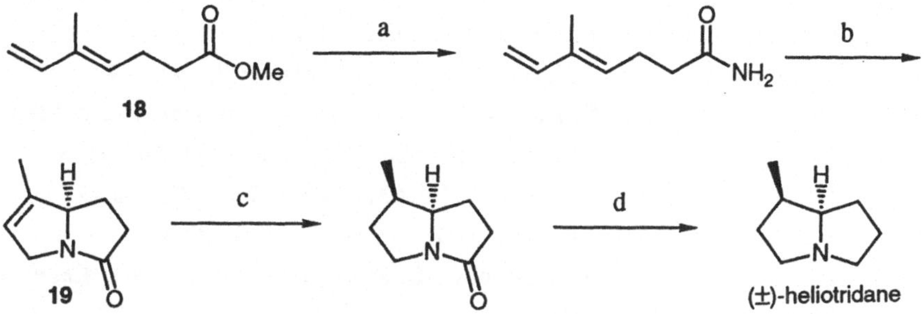

(a) NH_3, NaCN, MeOH, 75 %; (b) $Pd(OAc)_2$, $CuCl_2$ / O_2, THF, 60 °C, 85 %; (c) PtO_2, H_2, EtOH, 95 %; (d) $LiAlH_4$, ether, 89 %.

palladium-catalyzed cyclization afforded **19**. Catalytic hydrogenation and reduction of the carbonyl group completed the synthesis.

Acknowledgments. I wish to express my sincere appreciation to my collaborators, whose names appear in the references, for their efforts in exploring the chemistry outlined in this review. Financial support from the Swedish Natural Science Research Council and the Swedish Research Council for Engineering Science is gratefully acknowledged.

References and Notes

[1] J. Sauer, R. Sustman, *Angew. Chem.* **1980**, *92*, 773; *Angew. Chem. Int. Ed. Engl.* **1980**, *19*, 779.

[2] (a) M. Balci, *Chem. Rev.*, **1981**, *81*, 91. (b) G. Kaneko, A. Sugimoto, S. Tanaka, *Synthesis* **1974**, 876.

[3] S. M. Weinreb, R. R. Staib, *Tetrahedron,* **1982**, *38*, 3087.

[4] R. P. Vignes, J. Hamer, *J. Org. Chem.* **1974**, *39*, 849.

[5] G. E. Heasley, V. L. Heasley, S. L. Manatt, H. A. Day, R. V. Hodges, P. A. Kroon, D. A. Redfield, T. L. Rold, D. E. Williamson, *J. Org. Chem.* **1973**, *38*, 4109.

[6] (a) J. E. Bäckvall, S. E. Byström, R. E. Nordberg, *J. Org. Chem.* **1984**, *49*, 4619. (b) J. E. Bäckvall, J. O. Vågberg, R. E. Nordberg, *Tetrahedron Lett.* **1984**, *25*, 2717. (c) J. E. Bäckvall, J. O. Vågberg, *J. Org. Chem.* **1988**, *53*, 5695.

[7] (a) J. E. Bäckvall, A. K. Awasthi, Z. D. Renko, *J. Am. Chem. Soc.* **1987**, *109*, 4750. (b) J. E. Bäckvall, R. B. Hopkins, H. Grennberg, M. M. Mader, A. K. Awasthi, *ibid.* **1990**, *112*, 5161.

[8] (a) J. E. Bäckvall, J. E. Nyström, R. E.; Nordberg, *J. Am. Chem. Soc.*, **1985**, *107*, 3676. (b) J. E. Bäckvall, K. L. Granberg, R. B. Hopkins, *Acta. Chem. Scand.* **1990**, *44*, 492.

[9] (a) J. E. Nyström, J. E. Bäckvall, *J. Org. Chem.* **1983**, *48*, 3947. (b) J. E. Bäckvall,

S.E. Byström, J.E. Nyström, *Tetrahedron*, **1985**, *41*, 5761.

[10] (a) J. E. Bäckvall, H. E. Schink, Z. D. Renko, *J. Org. Chem.* **1990**, *55*, 826. (b) H. E.Schink, H. Pettersson, J. E. Bäckvall, *ibid.* **1991**, *56*, 2269.

[11] D. Tanner, M. Sellén, J. E. Bäckvall, *J. Org. Chem.* **1989**, *54*, 3374.

[12] (a) J. E. Bäckvall, P. G. Andersson, J. O. Vågberg *Tetrahedron Lett.* **1989**, *40*, 137. (b) J. E. Bäckvall, P. G. Andersson, *J. Am. Chem. Soc.* **1992**, *114*, 6374.

[13] J. E. Bäckvall, P. G. Andersson, *J. Am. Chem. Soc.* **1990**, *112*, 3683.

[14] J. E. Bäckvall, P. G. Andersson, *J. Org. Chem.* **1991**, *56*, 2274.

[15] J. E. Bäckvall, P. G. Andersson, G. B. Stone, A. Gogoll, *J. Org. Chem.* **1991**, *56*, 2988.

[16] (a) B. M. Trost, D. L. Van Vranken, *Angew. Chem. Int. Ed. Engl.* **1992**, *31*, 228. (b) G. Muchow, Dissertation, Université Aix Marseille III, Marseille, 1992.

[17] R. J. Kazlanskas, A. N. E. Weissfloch, A. T. Rappaport, L. A. Cuccia, *J. Org. Chem.* **1991**, *56*, 2988.

[18] This methodology was recently demonstrated on a related acyclic substrate, *E*-(2*S*,5*R*)-5-acetoxy-3-hexen-2-ol: H. E. Schink, J. E. Bäckvall, *J. Org. Chem.* **1992**, *57*, 1588.

[19] J.E. Bäckvall, R. Gatti, H.E. Schink, to be published.

[20] A. J. Pearson, Y. S. Lai, W. Lu, A. A. Pinkerton, *J. Org. Chem.* **1989**, *54*, 3882.

[21] A. D. Elbein, R. J. Molyneux, in "*Alkaloids: Chemical and Biological Perspectives*", S W. Pellitier (Ed.), Wiley: New York, 1987, vol 5, Chapter 1.

[22] T. Högberg, P. Ström, M. Ebner, S. Rämsby *J. Org. Chem.* **1987**, *52*, 2033.

[23] P. G. Andersson, J. E. Bäckvall, *J. Am. Chem. Soc.* in press.

New Developments in Zinc-mediated Organic Synthesis

Gerard van Koten*, Elmo Wissing, Henk Kleijn and Johann T.B.H. Jastrzebski

Department of Metal-Mediated Synthesis, Debye Research Institute, University of Utrecht, 3584 CH UTRECHT, The Netherlands

Summary

Results of current studies on mechanism of the reactions of diorganozinc compounds with 1,4-disubstituted-1,4-diaza-1,3-butadienes are discussed. Independent syntheses of some of the proposed cationic and neutral (radical) organozinc intermediates are described as well as the unique reactivity of these organozinc species which leads to selective synthesis of *e.g.* β-lactams, 2- and 3-pyrrolidinones and indolizines.

Introduction

Several years ago we discovered that dialkylzinc compounds (ZnR_2) react very selectively with 1,4-dihetero-1,3-butadienes such as the α-diimine, *t*-BuN=CHCH=N*t*-Bu (abbreviated as *t*-BuDAB) to either the N-alkylated product, *t*-Bu(R)NCH=CHN(ZnR)-*t*-Bu (R = primary alkyl group), or the C-alkylated product, *t*-Bu(RZn)NCH(R)CH=N*t*-Bu (R = tertiary alkyl group). In particular the finding that with primary alkyl groups (ethyl as well as the higher homologs but not for methyl, see below) selective N-alkylation occurred was unanticipated and prompted extensive further research in the Utrecht group. Some preliminary accounts on the results of our mechanistic studies have appeared[1-5]. Our present view is summarized in Chart 1 showing the two possible mechanisms that we use as working hypotheses in our current studies. As the synthesis and reactivity studies of some of the proposed intermediate zinc compounds shown in Chart 1 will be the subject of this paper, these two mechanistic proposals will be outlined first in some detail.

In both mechanisms the reaction starts with the formation of the 1:1 complex R_2Zn*t*-BuDAB which is followed by selective R-shift from Zn to either the imine-N or -C atom. The formation of R_2Zn*t*-BuDAB has been substantiated both by NMR-studies and the determination of the solid state structure of Me_2Zn*t*-BuDAB. The subsequent selective R-shift is different in both mechanisms and this point is the subject of further study.

The first mechanistic proposal, see Chart 1 **A**, comprises an intramolecular SET-step taking place in the initially formed, tetracoordinate R_2Zn*t*-BuDAB-primary complex which leads to the formation of a radical pair, [R$^{\cdot}$(RZn*t*-BuDAB)$^{\cdot}$], in a solvent cage. For this radical pair at least three further reaction pathways are available, which all three are supported by the

Dedicated with all best wishes to Prof. H. tom Dieck for his interest in our work and fruitfull discussions about DAB (DAD) chemistry on the occasion of his leave from the University of Hamburg.

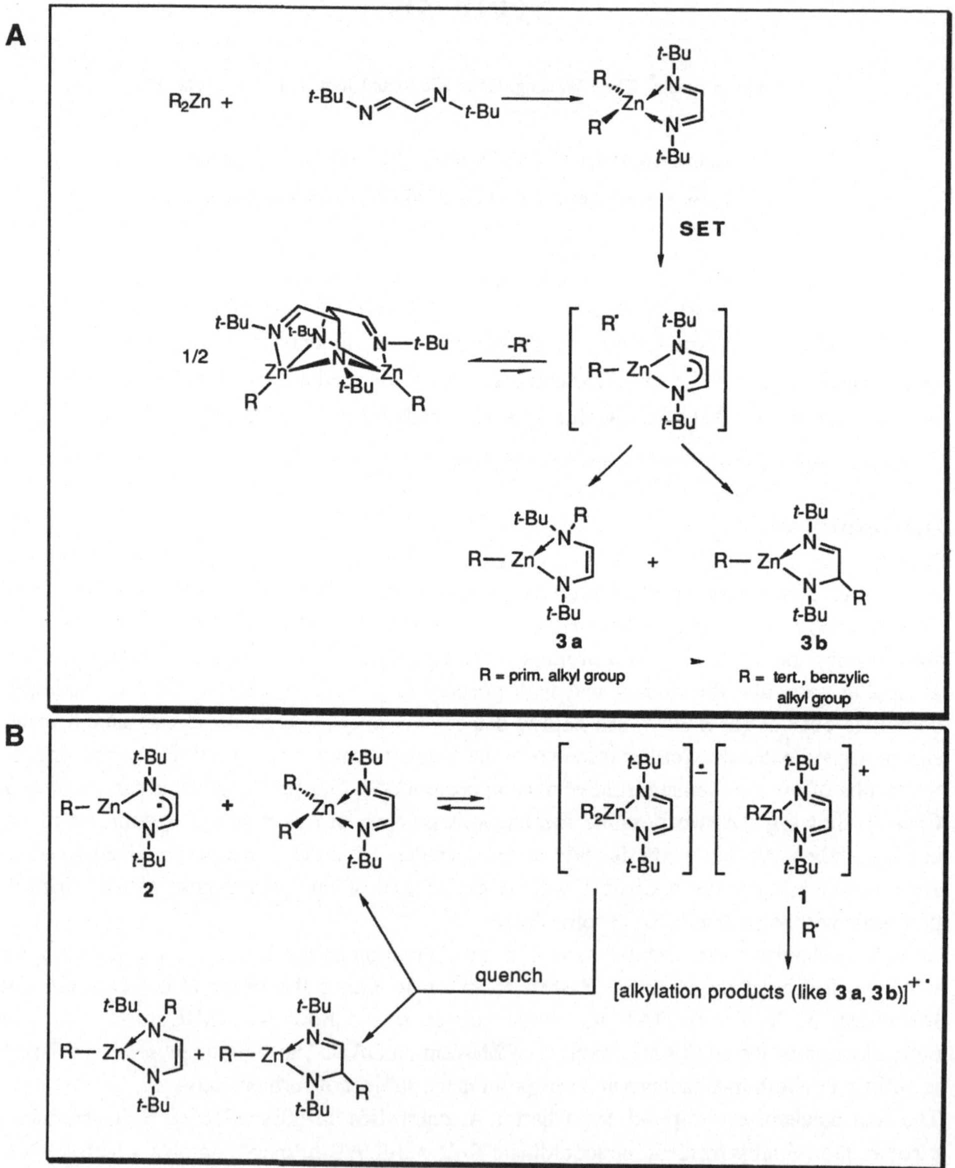

Chart 1

isolation of the respective organozinc products; *i.e.*, N- or C-alkylation through radical collaps as well as (RZn*t*-BuDAB)-radical formation via R-radical escape. The latter paramagnetic organozinc radical has been isolated, as will be discussed below, and exists in solution in equilibrium with its dimeric, diamagnetic counterpart (RZn*t*-BuDAB)$_2$. The second mechanistic proposal, see Chart 1 **B**, diverts from the first one at the stage of the formation of the radical pair and has as an essential subsequent step the formation of an ion-pair consisting of an (R$_2$Zn*t*-BuDAB)-radical anion and a (RZn*t*-BuDAB)-cation. This ion-pair is formed in an electron-transfer reaction between the (RZn*t*-BuDAB)-radical (formed by R-radical escape) and the initial tetrahedral (R$_2$Zn*t*-BuDAB)-complex. Product formation then occurs by attack of the (nucleophilic) R-radical on the (RZn*t*-BuDAB)-cation to give either the C- or N-alkylated products. An additional piece of information, which is crucial for the mechanistic discussion, is that products directly originating from reactions of the R-radical, *i.e.*, R$_2$ via radical-dimerization, RH via solvent proton abstraction and alkane (RH)/alkene (R–H) mixtures via radical-disproportionation, are not observed. Moreover, the complete mass-balance of the reaction of dihexylzinc with *t*-BuN=CHPy-2 (abbreviated as *t*-BuPyca) was established, see scheme 1.This quantitative analysis confirmed the occurrence of both C-alkylation, reduction as well as radical escape and indicated the existence of a radical disproportionation reaction between the HexZn*t*-BuPyca-radical and the Hex-radical (Note that for *t*-BuPyca N-alkylation is blocked). Each step in the mechanistic proposal for the Hex$_2$Zn/*t*-BuPyca reaction could be studied separatedly. Also, the formation of hexane and hexene in a 1:1 molar ratio could be unambiguously established.

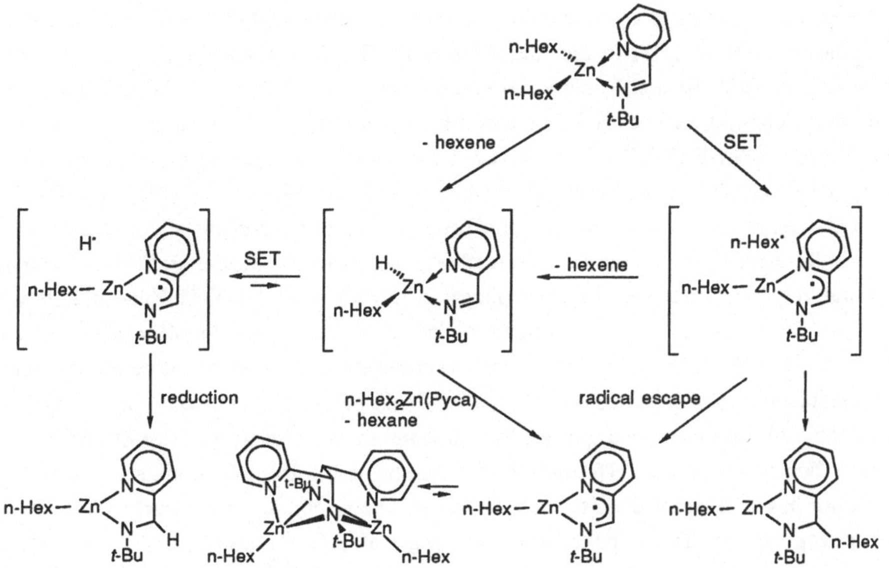

Scheme 1

These results indicate that in both mechanistic proposals of Chart 1, the kinetics of the separate steps play a crucial role. Because the organozinc- and R-radical separate by diffusion from the solvent cage free R-radical formation is possible and likely. However this requires that subsequent quenching-steps with the organozinc-*t*-BuDAB intermediates to give the recombination products are at least 10^2 faster then any direct reaction of the R-radical itself. Recently several of the proposed intermediates in both mechanistic proposals have been synthesized independently and their reactivity has been studied. The results of these studies do not only support our understanding of these fascinating reactions but also has led to synthetically interesting spin-offs; *i.e.*, the finding of novel routes for the enantioselective synthesis of all four enantiomers of 3,4-disubstituted 2-azetidinones [6] (β-lactam building blocks), 2-and 3-pyrrolidinones [7], indolizines [8], and functionally-substituted β-amino-imines [7] as well as the use of the N-alkylated products, which in fact are β-amino zinc-enamines, in aldol condensation reactions [9]. It is some of the latter research that will be the subject of this paper. For each section reference is given to one of the proposed intermediates in Chart 1.

Discussion

Synthesis and reactivity of the MeZn*t*-BuDAB-cation (see intermediate **1**)

An important reason to formulate an alternative mechanism for the initially proposed one was the observation by ESR of an organozinc radical-anion species $(R_2Znt\text{-BuDAB})^{\bullet -}$ in solutions of R_2Zn*t*-BuDAB in THF. In fact the RZn*t*-BuDAB-cation would be an obvious counter-cation for this radical-anion. It appeared that the complexes (RZn*t*-BuDAB)OTf (R = Me or 2,6-dimethylphenyl), can easily be prepared by reacting the corresponding 1:1 coordination complexes R_2Zn*t*-BuDAB (R = Me or 2,6-dimethylphenyl) with one equivalent of trifluoromethanesulfonic acid (HOTf). Best results in the synthesis of (MeZn*t*-BuDAB)OTf were obtained when a solution of Me_2Zn*t*-BuDAB in diethyl ether was added to a solution of HOTf in the same solvent. This procedure affords the off-white (MeZn*t*–BuDAB)OTf complex quantitatively. The first step in the reaction is the formation of OTf_2Zn*t*-BuDAB which subsequently reacts via a disproportionation reaction with added Me_2Zn*t*-BuDAB to give (MeZn*t*-BuDAB)OTf. Since the OTf-anion is known for its low coordinating ability and lack of nucleophilic character, it seemed likely that (MeZn*t*-BuDAB)OTf, could be a novel three-coordinate cationic organozinc species. The structure in the solid state of (MeZn*t*-BuDAB)OTf showed, however, that the OTf-anion is still coordinated to the zinc atom, see Figure 1. The zinc atom has a severely distorted tetrahedral geometry containing a nearly flat ZnNCCN-chelate ring with the Zn-Me bond making an angle of 130.4° with this plane, while the OTf anion binds to Zn via a slightly elongated Zn-O bond. ^{19}F NMR studies of (MeZn*t*-BuDAB)OTf in THF and benzene pointed to the existence of (solvated) (MeZn*t*-BuDAB)-cations in THF.

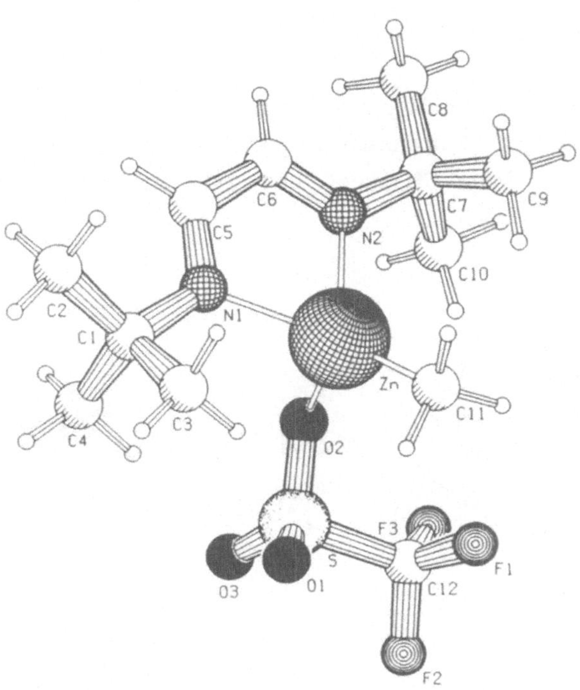

Figure 1. X-Ray structure of (MeZn*t*-BuDAB)OTf.

(MeZn*t*-BuDAB)OTf shows an interesting and selective ring-opening reaction with the cyclopropane derivative 1-trimethylsiloxy-1-ethoxycyclopropane, see eqn.1. After elimination of Me_3SiOTf and subsequent hydrolysis of the initially formed MeZnOEt-complex, the 2-pyrrolidinone derivative (**4**) is formed quantitatively. This reaction involves prior C-alkylation followed by an intramolecular nucleophilic substitution reaction of the amido function at the ethoxycarbonyl group. It is of interest to compare this result with the reactions of α-, β-, and γ–heteroatom substituted dialkylzinc compounds with *t*-BuDAB discussed later, see eqn. 6 and 7.

1 + 1-OEt-1-$OSiMe_3$-cyclopropane $\xrightarrow{-Me_3SiOTf}$ [Me–Zn(*t*-BuDAB-alkyl, $CH_2CH_2C(O)OEt$)] $\xrightarrow{H_2O}$ **4** (1)

Synthesis and reactivity of RZn*t*-BuDAB-radicals (see intermediate **2**)

The RZn*t*-BuDAB-radicals can be prepared in almost quantitative yield from the reaction of K(*t*-BuDAB) with RZnX or by thermal decomposition of the 1:1 coordination complex R_2Zn*t*-BuDAB (*e.g.*, for R = Me or CH_2SiMe_3). In the latter reaction the yield of the organozinc-radical is about 50%, see eqn. 2.

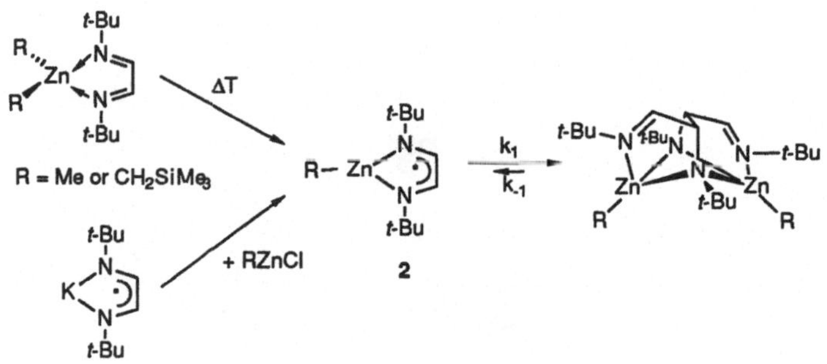

The organozinc radicals are paramagnetic species which have a very clear ESR spectrum, see Figure 2 A. It reveals that the species itself is planar with a trigonal coordination geometry of the zinc atom. From the ^1H and ^{14}N ESR hyperfine structure "experimental" spin-populations of φ_N and φ_{CH} of 0.3 and 0.2 could be calculated. The RZn*t*-BuDAB-radicals can be identified as the neutral systems $(RZn)^{+\cdot}$(*t*-BuDAB)$^{\cdot -}$. The α-proton splitting is caused by the hyperconjugation between σ(C-H) bonds of the zinc-alkyl substituent and the π^*(*t*–BuDAB) orbital in the trigonal planar arrangement at zinc [3].

The RZn*t*-BuDAB-radicals are in equilibrium in solution with the neutral dinuclear species (RZn*t*-BuDAB)$_2$, see eqn. 2, with can be isolated as (diamagnetic) solids. The structure in the solid state of two of such novel species have been established, *i.e*, of (EtZn*t*-BuPyca)$_2$, see scheme 1[5,10] and recently of (Me_3SiCH_2Zn*t*-BuDAB)$_2$ [11]. The latter structure shows that the two Me_3SiCH_2Zn*t*-BuDAB-radicals are coupled via a C-C bond, see figure 2B, which is significantly longer (1.62 Å) than a normal C(sp^3)-C(sp^3) single bond. This C-C bond formation is diastereoselective which may be due to the fact that only in the (*R*)(*S*) diastereoisomer both amido-nitrogens can bridge between the two zinc atoms. In this way, a new type of (NCCN)$_2$ bridging dianionic ligand is formed that renders both zinc atoms four-coordinated.

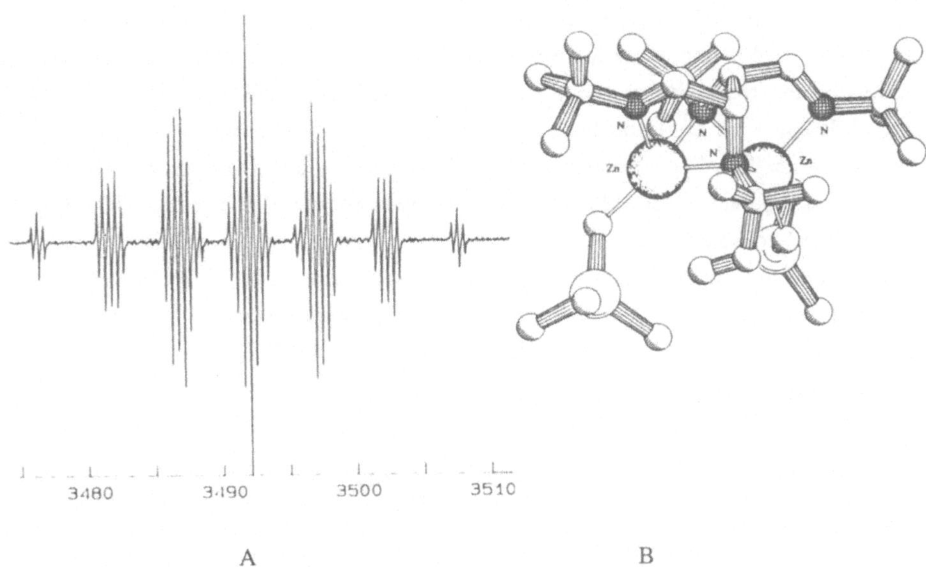

Figure 2. **A**, ESR spectrum of Me_3SiCH_2Znt-BuDAB-radical. **B**, X-ray structure of $(Me_3SiCH_2Znt\text{-BuDAB})_2$.

The concentration of the persistent RZn*t*-BuDAB-radical in solutions of these dinuclear species is small. Dinuclear $(EtZnt\text{-BuPyca})_2$ even has to be heated to 60 °C in benzene to give an ESR-detectable amount of the radical. The absence of line broadening in the NMR spectra of the dinuclear compound indicates that the rates, k_1 and k_{-1} (cf. eqn.2), are small as compared with the rate of the NMR-experiment. The equilibrium between the RZn*t*-BuDAB-radical and the dinuclear species $(RZnt\text{-BuDAB})_2$ was also established by the observation that upon mixing of two different, symmetric dinuclear species (dimer and dimer') in solution a third, unsymmetric dinuclear species was formed. The three species were found by NMR to be present in a 1:1:2 statistical ratio, see eqn. 3.

dimer ⇅ R–Zn(*t*-BuPyca) + dimer' ⇅ R–Zn(*t*-BuDAB) $\underset{k_{-1}}{\overset{k_1}{\rightleftharpoons}}$ unsymmetric dinuclear species (3)

The equilibrium between the dinuclear species and their mononuclear organozinc radical counterparts is temperature-dependent. Upon raising the temperature the equilibrium shifts in the direction of the radical species, as indicated by an enhancement of the intensity of the ESR signal and deepening of the colour of the solution. It appeared that at higher temperature a synthetically useful concentration of the radical in solution is present. We therefore investigated the reactivity of these slightly nucleophilic organozinc radicals towards electrophiles such as allyl and benzyl bromide, ethyl 2-bromopropionate and ethyl 3-chloropropionate. The dinuclear species (AlkylZn*t*-BuDAB)$_2$ could not be used for reasons outlined in a separate paper [11]. However, benzene solutions of (PhZn*t*-BuDAB)$_2$ and (EtZn*t*-BuPyca)$_2$ are stable up to 80 °C and have been used for these reactions. The reactions were completed in about 6 hours and resulted, for example, in the case of (PhZn*t*-BuDAB)$_2$ in the formation of a quantitative amount (50%) of the coordination complex PhZnX(*t*-BuDAB) (X = Cl or Br) and 50% of the C-alkylated products PhZn(*t*-BuNC(H)RC(H)=N*t*-Bu (for R = allyl or benzyl), see eqn. 4.

RX

BrCHMeCOOEt

$ClCH_2CH_2COOEt$

(4)

The reactions with the α- and β-substituted bromo- or chloro-esters gave after 16 hours products originating from C-alkylation followed by intramolecular cyclization. This results in the formation of zinc coordination complexes which after hydrolysis gave the free products, *i.e.*, a 3-methyl-4-tert-butylimino- (or 2-pyridy)l-2-azetidinone (d.e.'s between 90 and 60%) and a 2-pyrrolidinone, see eqn. 4. The actual yield of the alkylated products, which in one cycle is 50%, can be increased beyond that because the organozinc-radical can be reformed by treatment of the Zn(X)Ph(*t*-BuDAB) complexes with one equivalent of potassium. It must be noted that the compound with R = benzyl can also be obtained via a much shorter route involving the direct reaction of $(PhCH_2)_2Zn$ with *t*-BuDAB. Still, these reactions are of interest because the direct reaction of *t*-BuDAB with the corresponding diorganozinc compounds containing a β-ester functionality affords the N-alkylated product from which the 3-pyrrolidone is formed by intramolecular cyclization. Accordingly the reactions in eqns.(1) and (4) are complementary and point to the fact that in the direct reaction the radical recombination most probably is an inner-sphere process.

Syntheses and reactivity of β-amino zinc enamines (see intermediate 3)

As shown in Chart 1 the reactions of primary dialkylzinc compounds with *t*-BuDAB give rise to a quantitative and regioselective formation of N-alkylation products. These reactions seem to be general for all R'DAB systems in which R' = alkyl, *e.g.*, *i*-Pr, cyclo-Hex, neo-Pentyl, but isolation of the pure alkylated products in these cases was more difficult than with *t*-BuDAB itself. This is due to the general instability of the endiamines/aminoimines obtained after hydrolysis of the organozinc products. For this reason it is recommended to store these products as the organozinc precursors.

The β-amino zinc enamine obtained from, *e.g.*, the Et_2Zn/*t*-BuDAB reaction has been fully characterized. It is a distillable oil, which according to molecular weight determinations and NMR studies has a monomeric structure. The zinc atom most likely has a trigonal planar coordination by binding to the ethyl group, a neutral tertiary N atom and the amido N*t*-Bu anion. In particular the latter structural aspect is interesting because this amido grouping is part of an aza-allyl system when the C=C-N(*t*-Bu) as a whole is considered. Computational studies [2,12] on a model of these N-alkylated products, MeZnN(H)MeCH=CHNH, have shown that the Zn-N bond is short (1.88 Å) while a certain delocalization of charge into the C=C bond seems possible. This will cause a somewhat lower nucleophilicity of the amido N-atom. The calculated C-C distance (1.35 Å) indicates a C-C double bond whereas the C-N(amido) distance (1.47 Å) lies between single and double bond values. Together with an almost planar chelate ring this supports a three-center delocalization. The Zn-N(amino) bond (2.24 Å) indicates a coordinative bond to an amine which renders a description of these complexes in terns of β-amino zinc enamine complexes with contribution from a carbanionic resonance structure with negative charge on C_β feasible. Finally these calculations indicate the presence of a HOMO that is delocalized almost over the entire chelate ring with the highest AO coefficients on the amido-N and β-C atoms [3].

We currently use this regioselective N-alkylation method to introduce functionalized alkyl groups in RDAB-moieties via $Zn[(CH_2)_nXR']_2$ in which n = 1, 2 or 3, while XR' = OMe, SEt, OCH_2Ph, NMe_2, COOEt. These studies have led to the development of a new synthetic route for the synthesis of highly functionalized Z-1,2-diaminoethenes and 1-imino-2-aminoethanes. Some relevant examples are shown in Table 1 and eqn. 5. Also in this case the organic products obtained by hydrolyses of the organozinc derivatives had only limited stability and in our further chemistry the products are set free just before use or are reacted in the next synthetic step as the, stable, zinc precursor complexes.

The intramolecular coordination of the donor atoms in the functionalized alkylzinc reagents did not influence either the rate of reaction or the product formation. However, the mixture of

$Zn[(CH_2)_3NMe_2]_2$ with t-BuDAB had to be heated at 70 °C for 48 h in order to obtain complete transfer of the alkyl group from zinc to the DAB ligand. This reaction resulted in a product mixture of 68% N- and 21% C-alkylated products together with two unidentified minor products. A plausible explanation of this result is based on the fact that the Zn-NMe_2 coordinate bond is much stronger than the Zn-OR and Zn-SR bonds, respectively. The results indicate that the intermolecular Zn-N(imine) coordination can not compete with the intramolecular Zn-N(amine) coordination. Accordingly the higher reaction temperature and longer reaction time needed in the reaction of the NMe_2-substituted compound indicate that Zn-N bond dissociation is the rate-determining step.

(5)

Table I. Results of the alkylation reactions of R_2Zn compounds with *t*-BuDAB

ZnR_2 compound	N-alkylation (%)[a]	C-alkylation(%)
$Zn[(CH_2)_3OMe]_2$	100	0
$Zn[(CH_2)_3OCH_2Ph]_2$	100	0
$Zn[(CH_2)_4OMe]_2$	100	0
$Zn[(CH_2)_3SEt]_2$	100	0
$Zn[(CH_2)_3NMe_2]_2$	68	21
$Zn(CH_2CMe_3)_2$	0	100
$Zn(CH_2CMe_2Ph)_2$ [c]	0	100
$Zn(CH_2CH_2COOEt)_2$	100	0
$Zn(CHMeCH_2CH_3)_2$	5	95
$Zn[CHMe(CH_2)_2OMe]_2$	3	97
$Zn(CH_2Ph)_2$	5	95
$Zn(CH_2C_6H_4NMe_2\text{-}2)_2$ [c]	0	100
$Zn(C_6H_4CH_2OMe\text{-}2)_2$	1:1 coordination complex[d]	
$Zn(C_6H_4CH_2NMe_2\text{-}2)_2$	no reaction[e]	

[a] Percentage N- and C-alkylation after hydrolysis, determined by ^1H NMR integration of characteristic proton signals. [b] Complete conversion after heating for 48 h at 70 °C. [c] N- and C-alkylation percentages of the crude reaction product. [d] A thermally stable 1:1 coordination complex was isolated. [e] The starting materials were recovered unchanged.

Two other primary dialkylzinc compounds, $Zn(CH_2CMe_3)_2$ and $Zn(CH_2CMe_2Ph)_2$ likewise reacted differently with *t*-BuDAB. Instead of N-alkylation these organozinc reagents gave at room temperature >99% yield of C-alkylated products. At -78 °C they produced a 95 :5 molar mixture of C- and N-alkylated products indicating that alkyl-group transfer occurs under kinetic control. It must be noted that the Me_3CCH_2 and $PhMe_2CH_2$-radicals have relatively long lifetimes and thus give the thermodynamically more stable C-alkylation product. A further interesting point is that these radicals do not rearrange during the transfer process which indicates that they stay within the coordination sphere of the organozinc radical (see Chart 1) and alkylate before any rearrangement to CMe_2CH_2Me and CMe_2CH_2Ph radicals can take place. The rate of neophyl rearrangement processes is known (762 s^{-1}) [13] which indicates that the C-alkylation process must be at least 10^4 times faster and the N-alkylation step at least another 10^2 times. Also the ortho-amino benzylzinc compound, $(2\text{-}Me_2NC_6H_4CH_2)_2Zn$, reacts as expected with *t*-BuDAB to give the C-alkylated product (see eqn 5). In the reaction of the diorganozinc compound, $Zn[(CH_2)_2COOEt]_2$, with *t*-BuDAB in a 1:1 molar ratio at first the dark red color of the corresponding 1:1 complex was observed. Subsequently this complex reacts further along the route outlined in eqn 6 to give the 3-pyrrolidinone derivative after heating of the solution. The first step in the formation of this interesting product is N-alkylation resulting in the formation of the zinc-enamine having a nucleophilic β-C-atom. An intramolecular nucleophilic attack of the enamine-C atom on the carbonyl carbon then leads to the formation of the 3-pyrrolidinone derivative. Simultaneously with this C-C bond formation, an ethoxide group migrates to the zinc atom leading to an alkylzinc-ethoxide moiety which is coordinated to the newly formed 3-pyrrolidinone compound.

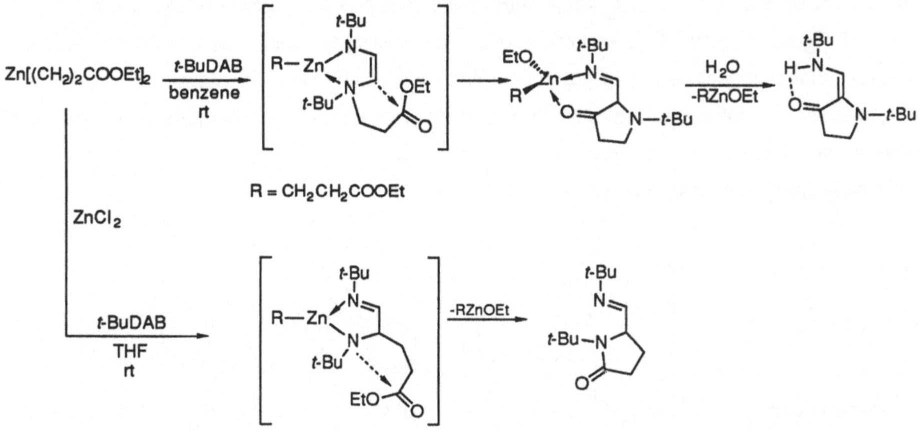

$Zn[(CH_2)_2COOEt]_2$ —(*t*-BuDAB, benzene, rt)→ [intermediate] → (EtO)(R)Zn-complex —(H_2O, -RZnOEt)→ product (6)

R = CH_2CH_2COOEt

$ZnCl_2$; *t*-BuDAB, THF, rt → [intermediate] —(-RZnOEt)→ product (7)

Interestingly, the selectivity of this reaction can be tuned in the direction of the formation of the

2-pyrrolidinone derivative, see eqn 7, by starting from a mixture of pure $Zn[(CH_2)_2COOEt]_2$ and the monoalkylzinc product. Instead of N-alkylation the functionalized alkylester grouping is transferred to the C-atom followed by intramolecular ring closure to the 2-pyrrolidinone product. At this point of our research we have no satisfactory explanation for this surprising result. The nucleophilicity of the β-amino zinc-enamines was also tested in several C-C coupling reactions with aldehydes. These reactions, which lead in a fast, exothermic reaction to the quantitative formation of organozinc aldolates, have been summarized in eqn 8. Most zinc aldolates are isolated as white, cystallizable solids, which exist as dimers in benzene solution. The C-C coupling occurs in the case of aryl aldehydes with high anti-diastereoselectivity, whereas in the case of alkyl aldehydes a less selective reaction was observed with syn/anti ratios amounting to 40/60.

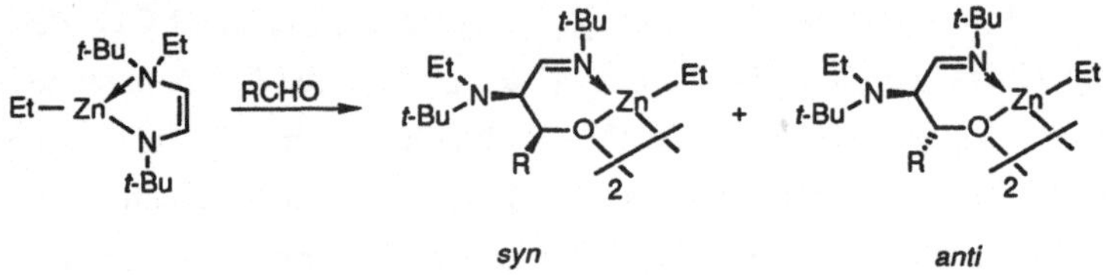

(8)

syn *anti*

Two further results, obtained with 4- and 2-pyridyl aldehyde, respectively, are particularly worth mentioning. The reaction with 4-pyridyl aldehyde and EtZn(Et)*t*-BuNCH=CHN*t*-Bu in an exact 1:1 molar ratio afforded quantitatively the anti-C-C coupled zinc-aldolate. As it is known that $ZnEt_2$ forms a 1:2 complex $ZnEt_2(Py)_2$ with pyridine the zinc-aldolate was reacted with a second equivalent of $ZnEt_2$. This reaction gave rise to the formation of an interesting coordination polymer in which two pyridine groups of different dinuclear zinc-aldolate units repeatedly coordinate to the $ZnEt_2$ molecules; for a part of the structure in the solid state see eqn 9 and figure 3 [8,9]. This structure is a nice example of self-assembly of different molecules, with the right match of stereochemical and electronic information, to a polymeric structure. In the case of the present structure each separate unit forms an enantiomeric pair, that is mirror-imaged in the next dimeric unit. It is this alternating sequence that is the reason for the relatively flat, rod-like stereochemistry of the polymeric structure.

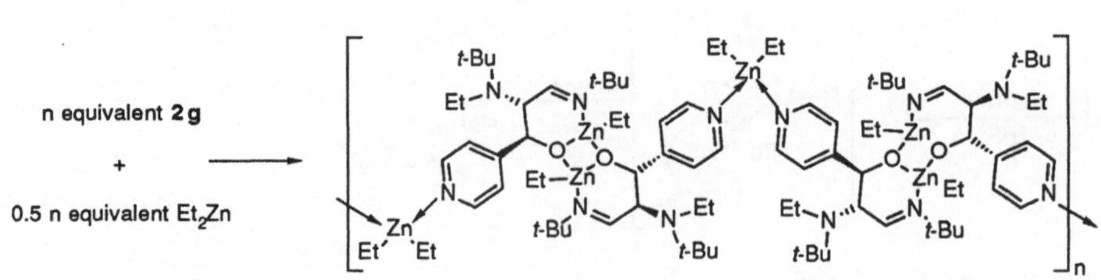

n equivalent **2g** + 0.5 n equivalent Et_2Zn (9)

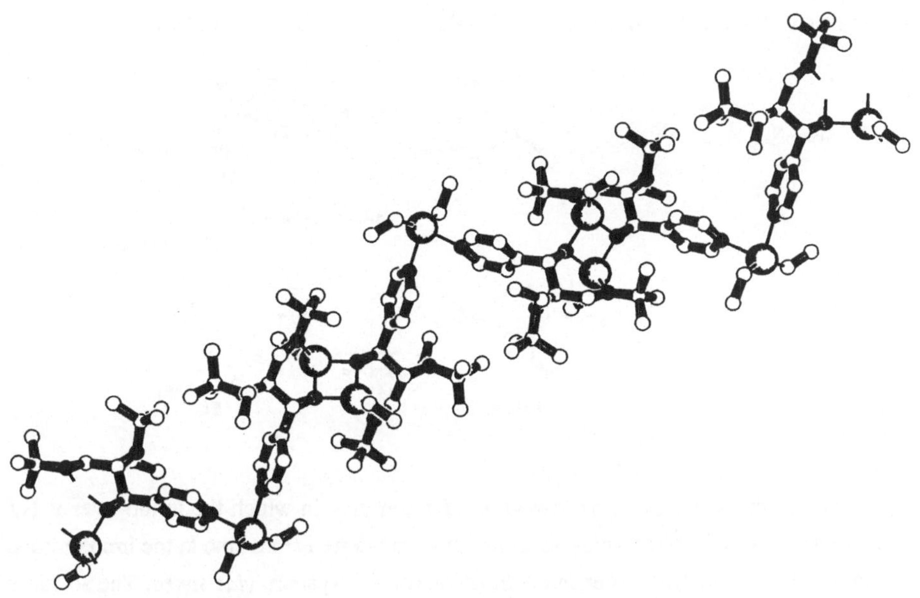

Figure 3. X-Ray structure of the polymeric C-C coupled zinc-aldolate $ZnEt_2$ complex.

The second surprising result was the outcome of the reaction of EtZn(Et)*t*-BuNCH=CHN*t*-Bu with 2-pyridyl aldehyde. This reaction cleanly affords indolizine product shown in eqn 10.

t-Bu, Et, N, *t*-Bu, N, Zn, Et, O, N, 2 $\xrightarrow[\text{-EtZnOH}]{\Delta T}$ *t*-Bu, N, N, *t*-Bu–N, Et $\xrightarrow[\text{72 h}]{\Delta T}$ N, *t*-Bu–N, H, Et, N, *t*-Bu (10)

Finally, an important further line of research originated from these studies is the use of β-aminozinc enolates for the stereoselective synthesis of *cis*- and *trans*-β-lactams. The initial discovery of the usefulness of these zinc enolates emerged from the reaction of iminoester (**5**) and diethylzinc in a 1 : 1 molar ratio. This reaction afforded rather unexpectedly the *trans*-β-lactam (**6**) in quantitative yield [14], eqn.11.

t-Bu–N, OMe, O (**5**) + Et_2Zn ⟶ 1/2 Et, MeO, *t*-Bu–N, O, O, N, *t*-Bu (**6**) + EtZnOMe + 1/2 Et_2Zn (11)

A mechanism explaining the formation of this β-lactam is outlined in Scheme 2.

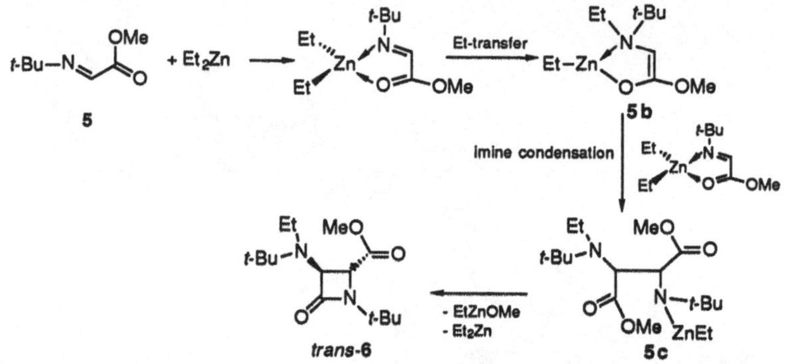

Scheme 2.

The first step in this reaction is the formation of a complex in which the imino-ester is N,O-chelate bonded to Et_2Zn, after which an Et group is transferred from zinc to the imine-nitrogen atom, comparable to the Et-transfer in the Et_2Zn/*t*-BuDAB system, *vide supra.* The so-formed ethylzinc enolate (**5b**) undergoes an aldol-like condensation reaction with unreacted iminoester, giving rise to the formation of the C–C coupled product (**5c**). Finally, the *trans*-β-lactam **6** is formed by elimination of EtZnOMe.

It is most probably the Et_2Zn-iminoester complex and not the iminoester itself that undergoes the condensation reaction with the ethylzinc enolate. This may be concluded from: *i*, the stoichiometry of the reaction shown in eq.11 (when the iminoester to Et_2Zn ratio is 2 : 1 only 50% of the β-lactam is obtained), and *ii*, our recent observation that in the condensation reaction of zinc enolates with imines derived from glycine ester only the $ZnCl_2$-glycine imine ester complex is reactive[15].

Based on these results we have developed a new general synthetic route to β-lactams, outlined in eq. 12. These reactions are carried out as simple one pot syntheses. First the appropriate ester is deprotonated with a base like LDA. The so-formed lithium enolate is transmetallated with $ZnCl_2$ and is finally reacted with the respective imine. After hydrolysis and work-up the desired β-lactams are obtained in high yield. Furthermore, making use of proper chiral subsituents (*e.g.* with R'''= (*R*)-2-phenylethyl) in the ester or the imine fragments the diastereo- and enantioselectivity of the reaction can be highly controlled[16].

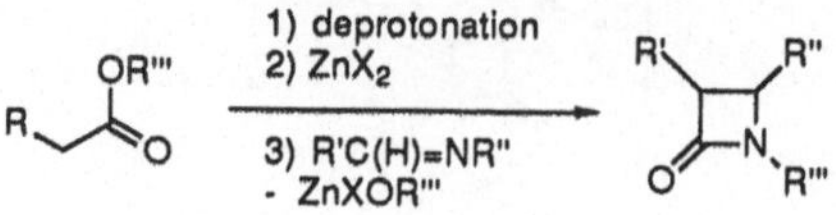

(12)

Concluding remarks

The present results show that a clear answer of the question, 'Are these zinc-mediated organic reactions occuring via a polar or a radical mechanism?' (the actual title of the lecture) at present is not possible. The mechanisms outlined in Chart 1 are both likely possibilities which need further study. Currently we use either special R_2Zn compounds (*e.g.* R = 4-hexenyl) or special R'-DAB (*e.g.* R' = cyclo-propyl) ligands which can exhibit particular reactivity patterns when intermediate stages of the reaction would comprise the formation of (free) radicals, cf. Chart 1. Furthermore, we are concentrating on the synthesis of some of the intermediates proposed in Chart 1B, *e.g.* the $\cdot[R_2ZnR'\text{-DAB}]^{\bullet -}.[RZnR'\text{-DAB}]^+$ complex. Despite the many questions that remain in this part of our research it is interesting to see the influence of the independent synthesis and subsequent reactivity studies of the proposed organozinc reaction intermediates on the direction of our organic synthetic research. Actually a number of new synthetic routes to both known and new organic products have been found. The mediating role of zinc in these reactions is also subject of further study in the Utrecht-group.

Acknowledgement

The work was supported by grants from the Netherlands Foundation for Chemical Research (SON) with financial aid from the Netherlands Organization for Scientific Research (NWO) to E.W.. The continuous interest and support by Drs. D.M. Grove and J. Boersma is gratefully acknowledged. We thank also Dr. A.L. Spek, W.J.J. Smeets and H. Kooijman (University of Utrecht) for their interest and x-ray structure determinations, and Prof. dr. W. Kaim and dr. M. Kaupp (University of Stuttgart) for their support with spectroscopic and theorectical calculations. Gist-brocades nv., Delft, The Netherlands, is acknowledged for their financial support of our research efforts in the field of the β-lactam chemistry.

References

[1] J.M. Klerks, J.T.B.H. Jastrzebski, G. van Koten, K. Vrieze, J. Organomet. Chem. 224 (1982) 107.

[2] G. van Koten in A. de Meijere, H. tom Dieck (Eds): Organometallics in Organic Synthesis, Springer-Verlag, Heidelberg, 1987, 277.

[3] M. Kaupp, H. Stoll, H. Preuss, W. Kaim, T. Stahl, G. van Koten, E. Wissing, W.J.J. Smeets, A.L. Spek, J. Am. Chem. Soc. 113 (1991) 5606.

[4] J.T.B.H. Jastrzebski, J.M. Klerks, G. van Koten, K. Vrieze, J. Organomet. Chem. 210 (1981) C49.

[5] G. van Koten, J.T.B.H. Jastrzebski, K. Vrieze, J. Organomet. Chem. 250 (1983) 49.

[6] F.H. van der Steen, H. Kleijn, G.J.P. Britovsek, J.T.B.H. Jastrzebski, G. van Koten, J. Org. Chem. 57 (1992) 3906.

[7] E. Wissing, H. Kleijn, J. Boersma, M.D. Janssen, A.L. Spek, G. van Koten, J. Org. Chem. to be published.

[8] E. Wissing, R.W.A. Havenith, J. Boersma, G. van Koten, Tetrahedron Let. accepted for publication.

[9] E. Wissing, R.W.A. Havenith, J. Boersma, W.J.J. Smeets, A.L. Spek, G. van Koten, J. Org. Chem. to be published.

[10] A.L. Spek, J.T.B.H. Jastrzebski, G. van Koten, Acta Cryst. C43 (1987) 2006.

[11] E. Wissing, S. van der Linden, J. Boersma, W.J.J. Smeets, A.L. Spek, G. van Koten, Organomettalics to be published

[12] M. Kaupp, H. Stoll, H. Preuss, J. Comp. Chem. 11 (1990) 1029.

[13] D.A. Lindsay, J. Lusztyk, K.U. Ingold, J. Am. Chem. Soc. 106 (1984) 7087.

[14] M.R.P. van Vliet, J.T.B.H. Jastrzebski, W.J. Klaver, K. Goubitz and G. van Koten, Recl. Trav. Chim. Pays-Bas 106 (1987) 132.

[15] H.L. van Maanen, J.T.B.H. Jastrzebski and G. van Koten, Tetrahedron Asymmetry submitted for puplication.

[16] F.H. van der Steen, H. Kleijn, G.J.P. Britovsek, J.T.B.H. Jastrzebski and G. van Koten, J. Org. Chem. 57 (1992) 3906.

Organometallics in Organic Synthesis via Radicals

Bernd Giese

Institut für Organische Chemie, Universität Basel, St. Johanns-Ring 19, CH-4056 Basel, Switzerland

Summary

Organometallics play an important role in radical chemistry. Thus, C,C-bond forming reactions via radical addition to π-bonds were developed using organomercury and organotin hydrides as mediators. Substitution of these metals by silicon offers new synthetic applications. Radical addition reactions of silanes with alkynes, alkenes, and ketones show remarkable stereoselectivities. Transition metals can also be used in radical chain reactions. *In situ* regeneration of the organometallic provides catalytic methods.

1 Introduction

Recently we have developed synthetic methods of C,C-bond formation *via* intermolecular radical addition to π-bonds using organomercury [1] and organotin [2] compounds as mediators [3].

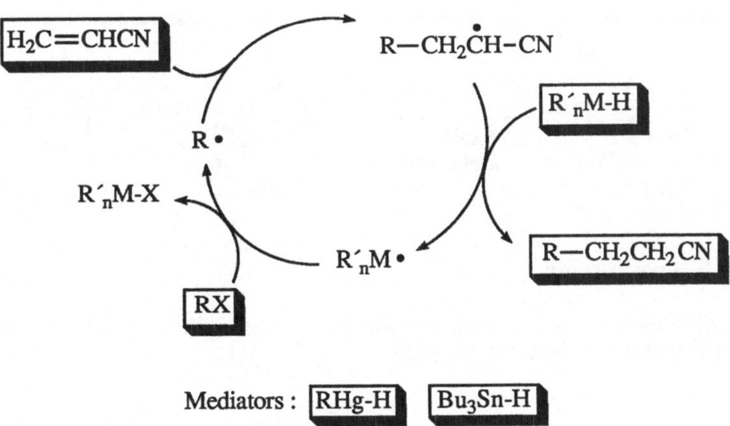

These methods became very popular but until the end of the last decade these methods suffered from the toxicity of these organometallics and from the low stereoselectivity of the intermolecular radical reactions. Thus, in our synthesis of malyngolide it was difficult to remove traces of tin and to carry out the radical reaction stereoselectively [4].

Malyngolide

C_9H_{19} ⟹ CO_2Me / Ph, O, O, I, C_9H_{19} ⟹ OH, C_9H_{19}

Sharpless → 55% ; 1. LiI 2. $PhCH(OMe)_2$ → 71%

$Bu_3Sn\bullet$, $-Bu_3SnI$ → ; CO_2Me → MeO_2C ; Bu_3SnH, $-Bu_3Sn\bullet$

70% ; 1. H_2 / Pd-C 2. KOH amberlyst 15 → 37% + 40%

2 Chiral Auxiliaries

Recent studies have now shown that chiral auxiliaries with C_2-symmetry induce high stereoselectivity also in radical reactions [5].

AUXILIARY CONTROL

R•   X–Y

With suitably substituted amines these reactions are completely stereoselective.

R•	CH_3 / CH_3 pyrrolidine	CH_2OCH_3 / CH_2OCH_3 pyrrolidine	bis-acetal (Ph)
t-C_4H_9 •	82 : 1	112 : 1	> 200 : 1
c-C_6H_{11} •	48 : 1	67 : 1	> 200 : 1
n-C_6H_{13} •	16 : 1	35 : 1	> 200 : 1

3 Tris(trimethylsilyl)silane

It is also possible to replace stannanes by silanes as mediators in these radical chain reactions if one uses tris(trimethylsilyl)silane [6], which is now easily available [7].

BDE (kcal/mol): 74 → Bu_3Sn-H; 90 → Et_3Si-H; 79 → $(Me_3Si)_3Si$-H

$$Me_3SiCl + SiCl_4 \xrightarrow[THF]{Li} (Me_3Si)_4Si \xrightarrow[THF]{MeLi} (Me_3Si)_3Si\text{-}H \quad 80\%$$

Tris(trimethylsilyl)silane is not only a beneficial mediator for C,C-bond forming reactions it also adds easily to π-bonds via radical chain reactions [8].

$$>C{=}O \xrightarrow{(TMS)_3SiH} H{-}\overset{|}{\underset{|}{C}}{-}OSi(TMS)_3$$

$$>C{=}C< \xrightarrow{(TMS)_3SiH} H{-}\overset{|}{\underset{|}{C}}{-}\overset{|}{\underset{|}{C}}{-}Si(TMS)_3$$

These reactions can occur with high stereoselectivity.

H—≡—X + $(TMS)_3Si$-H $\xrightarrow[-15^\circ C]{BEt_3 / O_2}$ $(TMS)_3Si$(H)C=C(H)X

90 %   Z : E ≥ 99 : 1

$-Si$ H X

Me (maleic anhydride) $\xrightarrow[-15^\circ C]{(TMS)_3SiH}$ $(TMS)_3Si$, Me (succinic anhydride)

90 %   Z : E = 99 : 1

H, CO_2Et, EtO_2C, Me $\xrightarrow[20^\circ C]{(TMS)_3SiH}$ $(TMS)_3Si$, CO_2Et, Me, H, EtO_2C, H

79 %   threo : erythro = 25 : 1

Whereas stereoselective reactions of alkynes and cyclic alkenes were expected [9], the high selectivities of flexible radicals, generated from alkenes, were somehow surprising. Selectivity and ESR studies have now shown that acyclic radicals can adopt preferred conformations [10].

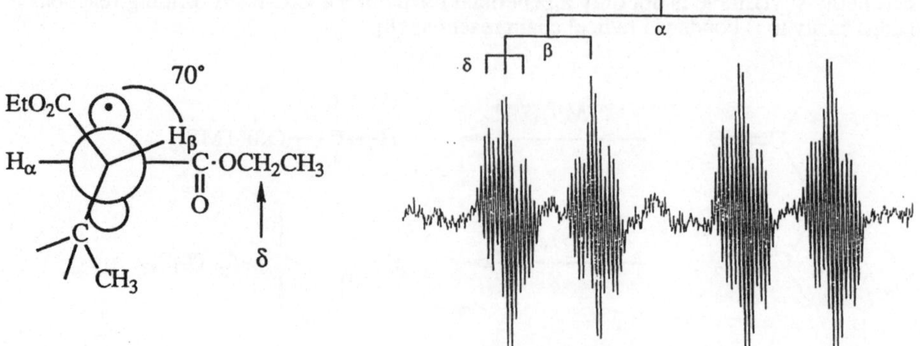

These preferred conformations are explained by allylic strain effects [11].

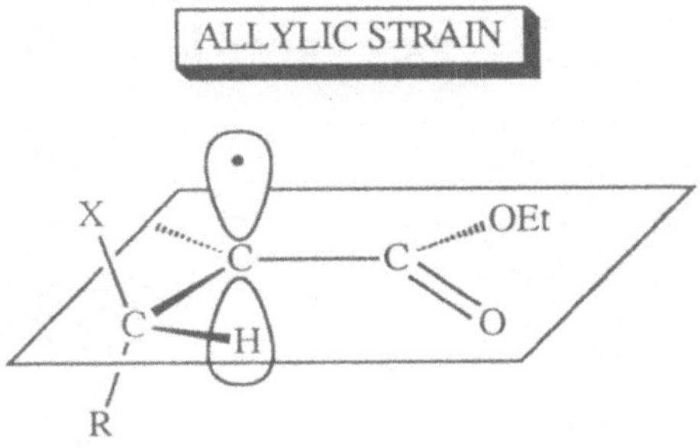

ALLYLIC STRAIN

The fast attack of Si radicals to the oxygen atom of carbonyl groups makes it possible to study the question whether Cram or Felkin-Anh rules can be applied also to radical chemistry.

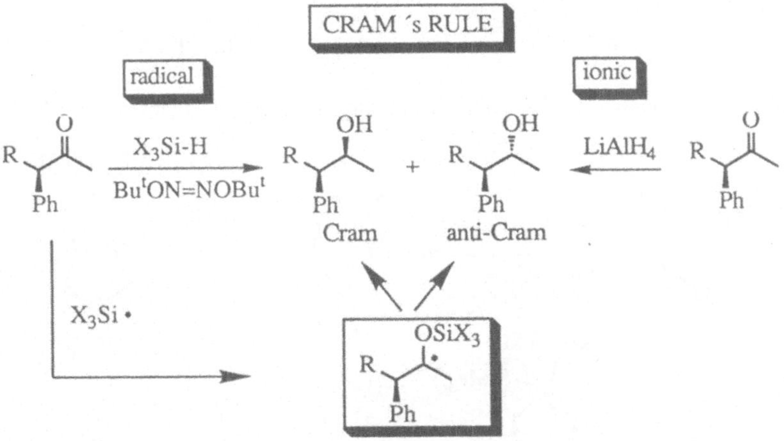

CRAM 's RULE

radical

ionic

$X_3Si\text{-}H$, $Bu^tON{=}NOBu^t$

$LiAlH_4$

Cram

anti-Cram

$X_3Si\bullet$

ESR-experiments prove that tris(trimethylsilyl)silyl radicals add cleanly to the oxygen of chiral ketones and lead to a preferred conformation [12].

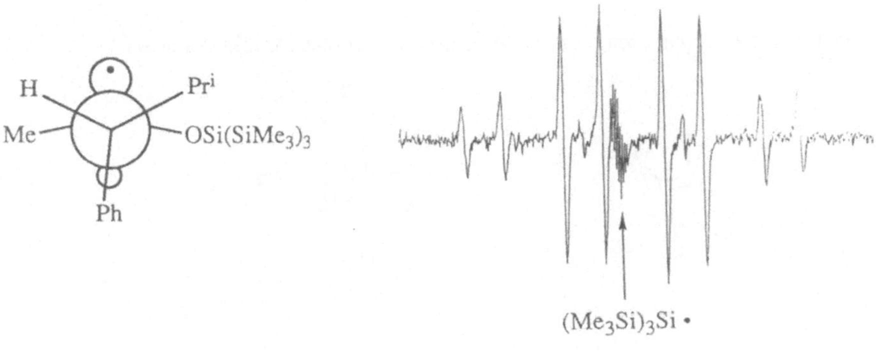

$(Me_3Si)_3Si\bullet$

Trapping experiments show that the stereoselectivities of ionic and radical reactions are comparable [12][13].

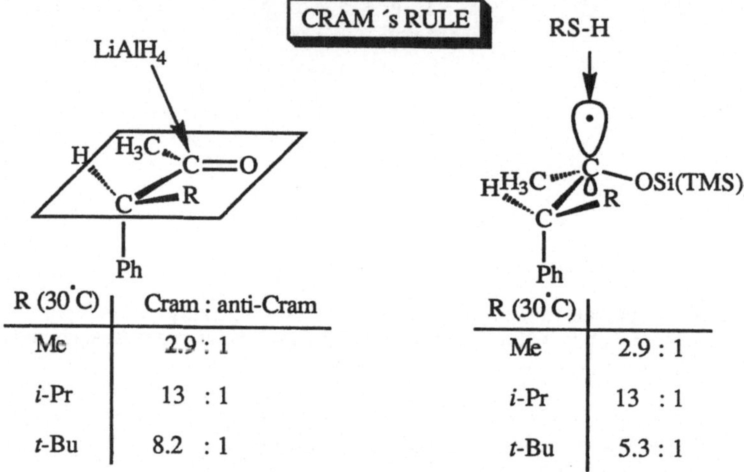

R (30°C)	Cram : anti-Cram
Me	2.9 : 1
i-Pr	13 : 1
t-Bu	8.2 : 1

R (30°C)	
Me	2.9 : 1
i-Pr	13 : 1
t-Bu	5.3 : 1

Organosilicon compounds can also act as radical traps in intramolecular reactions [14].

$(TMS)_3Si\text{-}H$ / 85°C, slow addition → Si(TMS)3 product (25 %) + TMS, TMS-Si bicyclic product (55 %)

TMS TMS Si SiMe3 … X

4 Transition Metal Complexes

Organoiron and organocobalt compounds can be used as sources for alkyl radicals.

Fe, OC, CO, R; R, Co, N, O, H, py

$$L_nM\text{—}R \xrightarrow{h\nu} L_nM\bullet + R\bullet$$

The radicals are formed *via* metal-carbon bond cleavage. After trapping of the alkyl radicals with alkenes a new metal-carbon bond is formed. Subsequent hydrolysis, reduction or metal hydride elimination yields the products [15].

R–CoLn + CH_2=C(OSiMe$_3$)(CO$_2$R) —hν→ R(H)C=C(OSiMe$_3$)(CO$_2$R) → R–CH$_2$–CO–CO$_2$R

R–CoLn ⇌(hν) R• •CoLn —(alkene-Y)→ R–CH$_2$–CH(•)Y •CoLn ⇌ R–CH$_2$–CH(Y)–CoLn

↓ - HCoLn

R–CH=CH–Y

Free Radicals

Rates, Selectivities

We have used this method in the biomimetic synthesis of KDO [13] and of C-glycosides [16].

HO–CH$_2$–CH(OH)–CH(OH)–CH(OH)–CHO + CH_2=C(OⓅ)(CO$_2^{\ominus}$) —enzyme→ HO–CH$_2$–CH(OH)–CH(OH)–CH(OH)–CH(OH)–CH$_2$–CO–CO$_2$H in vivo

→ KDO

AcO–CH$_2$–CH(OAc)–CH(OAc)–CH(OAc)–CH(OAc)–Co(dmgH)$_2$py + CH_2=C(OTMS)(CO$_2$Me) —hν, 72%→ AcO–CH$_2$–CH(OAc)–CH(OAc)–CH(OAc)–CH=CH–CO–CO$_2$Me in vitro

In trapping reactions with NO [17] or O_2 [18] carbon-nitrogen and carbon-oxygen bonds, respectively, are formed.

R-CoLn $\xrightarrow[\text{NO}]{h\nu}$ **R-NO**

NaCo(dmgH)$_2$py

Co(dmgH)$_2$py

80 %

NO / hν

H$_2$ / Pd-C

HOAc / Ac$_2$O

NOH

NHAc

75 %

76 %

Scheffold has shown that coenzyme B_{12} can be used as catalyst for radical C,C-bond formation [19]. These reactions occur *via* a radical chain reaction in which Co adopts the oxidation states +I, +II, and +III.

Catalysis

R-X

$-X^-$

$[Co^{III}]$–R

hν

$[Co^{II}]$ + R•

Y

Products

$[Co^{I}]^-$

e-

From the reduction potentials we concluded that "Co(Costa)" and "Co(tim)" complexes should be as efficient as coenzym B_{12}.

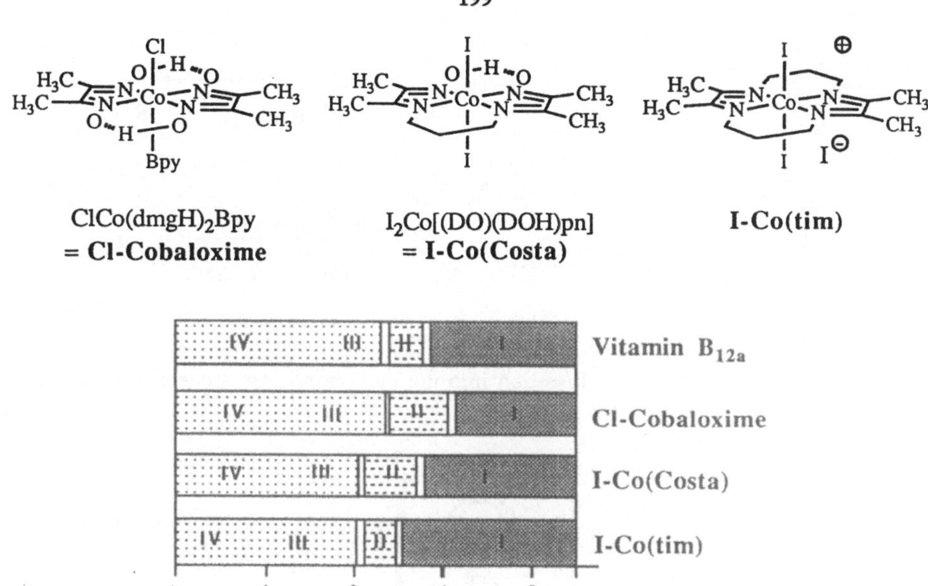

Addition and cyclization experiments proved this.

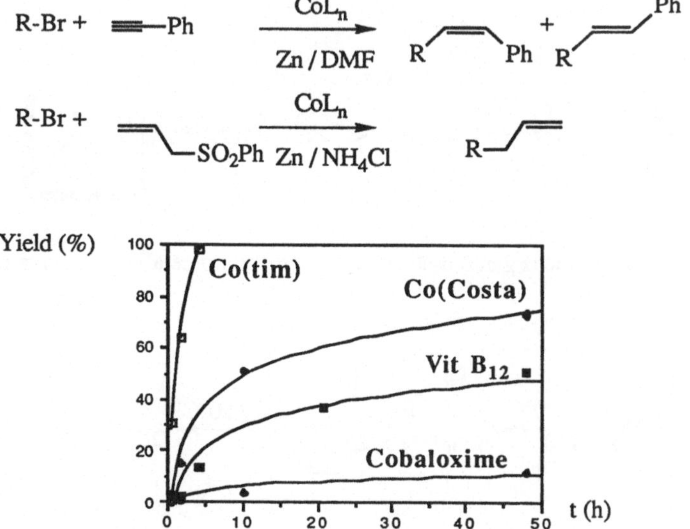

Depending on the conditions either addition or substitution products are formed [20].

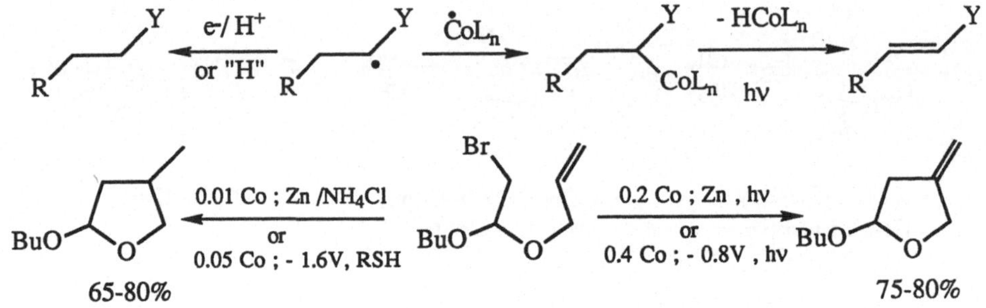

With iron carbonyl complexes CO-insertion into the Fe,C-bond can occur prior to radical formation [21].

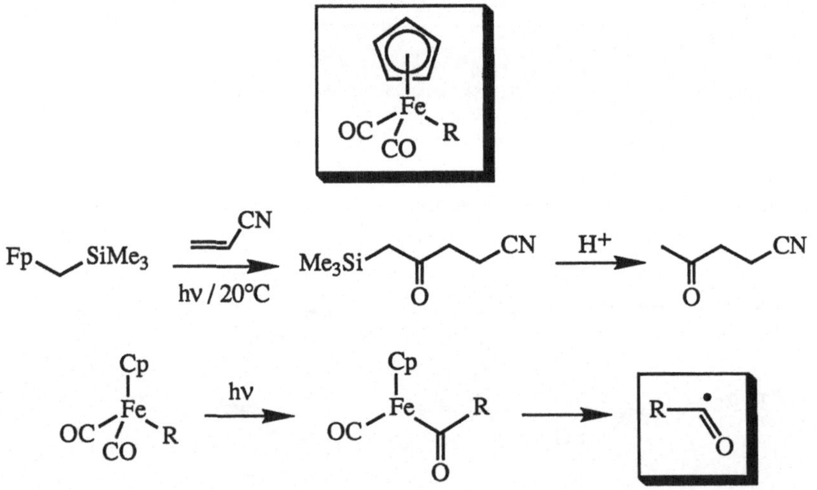

Adjacent olefinic bonds in the organic molecules lead to organometallic rearrangements prior to the radical reaction [22].

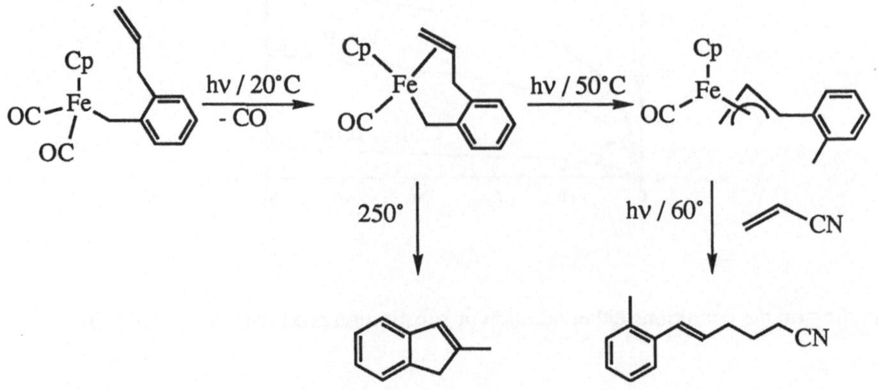

Acknowledgement

This work was supported by the Volkswagen-Stiftung and the Swiss National Science Foundation.

References

[1] B. Giese, J. Meister, *Chem. Ber.* **1977**, *110*, 2558.
[2] B. Giese, J. Dupuis, *Angew. Chem.* **1983**, *95*, 633.
[3] For reviews see: B. Giese, *Angew. Chem.* **1985**, *97*, 555; W.P. Neumann, *Synthesis* **1987**, 665; D.P. Curran, *Synthesis* **1988**, 489.
[4] B. Giese, R. Rupaner, *Liebigs Ann. Chem.* **1987**, 231.
[5] For a review see: N.A. Porter, B. Giese, D.P. Curran, *Acc. Chem. Res.* **1991**, *24*, 296.
[6] B. Giese, B. Kopping, C. Chatgilialoglu, *Tetrahedron Lett.* **1989**, *30*, 681.
[7] B. Giese, J. Dickhaut, *Org. Syntheses* **1991**, *70*, 164.
[8] K.J. Kulicke, B. Giese, *Synlett* **1990**, 91.
[9] B. Giese, *Angew. Chem.* **1989**, *101*, 993.
[10] B. Giese, W. Damm, F. Wetterich, H.-G. Zeitz, *Tetrahedron Lett.* **1992**, *33*, 1863.
[11] B. Giese, M. Bulliard, H.-G. Zeitz *Synlett* **1991**, 425.
[12] B. Giese, W. Damm, J. Dickhaut, F. Wetterich, S. Sun, D.P. Curran, *Tetrahedron Lett.* **1991**, *32*, 6098.
[13] B. Giese, B. Carboni, T. Göbel, R. Muhn, F. Wetterich, *Tetrahedron Lett.* **1992**, *33*, 2673.
[14] K.J. Kulicke, C. Chatgilialoglu, B. Kopping, B. Giese, *Helv. Chim. Acta* **1992**, *75*, 935.
[15] J. Hartung, J. He, O. Hüter, A. Koch, B. Giese, *Angew. Chem.* **1989**, *101*, 334; see also: B.P. Branchaud, M.S. Meier, Y. Choi, *Tetrahedron Lett.* **1988**, *29*, 167; G. Pattenden, *Chem. Soc. Rev.* **1988**, *17*, 361.
[16] A. Ghosez, T. Göbel, B. Giese, *Chem. Ber.* **1988**, *121*, 1807.
[17] A. Veit, B. Giese, *Synlett* **1990**, 166.
[18] J. Hartung, B. Giese, *Chem. Ber.* **1991**, *124*, 387.
[19] R. Scheffold, *Chimia* **1985**, *39*, 203.
[20] B. Giese, P. Erdmann, T. Göbel, R. Springer, *Tetrahedron Lett.* in press.
[21] B. Giese, G. Thoma, *Helv. Chim. Acta* **1991**, *74*, 1143.
[22] G. Thoma, B. Giese, *Helv. Chim. Acta* **1992**, *75*, 1123.

Acknowledgement

This work was supported by the Volkswagen-Stiftung and the Swiss National Science Foundation.

References

[1] [illegible] 197[illegible]
[2]–[9] [illegible]
[10] [illegible]
[11] [illegible] 1991 [illegible]
[12] [illegible]
[13] [illegible]
[14] [illegible]
[15] [illegible]
[16] [illegible]
[17] [illegible]
[18] [illegible] 1991, 324, 22[illegible]
[19] [illegible] 1985 [illegible]
[20] [illegible] in preparation
[21] [illegible] 1991 [illegible]
[22] [illegible] 1992, 75, [illegible]

α-Heteroalkenyl Metallate Rearrangements in Organic Synthesis

Philip Kocieński

Department of Chemistry, The University, Southampton, SO9 5NH, U.K.

Summary

The scope, stereochemistry and mechanism of 1,2-metallate rearrangements of borates and cuprates derived from 5-lithio-2,3-dihydrofuran and 6-lithio-2,4-dihydro-2H-pyrans is compared with borates and cuprates derived from α-metallated enol carbamates. Depending on the metallate complex and the substrate, 1,2-alkyl shifts can take place by several mechanisms.

1. Introduction

The intermolecular transfer of a nucleophilic carbon ligand from a metal to an electrophilic carbon forms the widest class of C-C bond forming reactions in common use today. The reaction of Grignard reagents with carbonyls and organocuprates with haloalkanes are obvious and ubiquitous examples of addition and substitution reactions which form the bedrock of organic synthesis. Far less common are intramolecular variants in which both the nucleophilic and the electrophilic partners are bound to the metal. When the metal bears a full negative charge such reactions may be defined as 1,2-metallate rearrangements and examples are known of migration to sp^3 [1] sp^2 [2], and sp-hydridised carbon [3] as exemplified in Scheme 1. A significant advantage of these reactions is the retention of a carbon-metal bond in the product which can then be used for further synthetic transformations. In the following review the scope, stereochemistry and mechanism of 1,2-α-heteroalkenylborates and -cuprates will be considered along with a few synthetic applications.

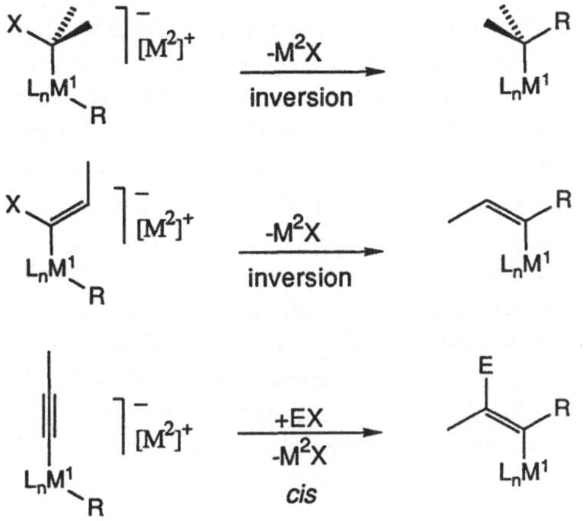

X, L_nM^1, R, $[M^2]^+$, $-M^2X$, inversion, R, L_nM^1

X, L_nM^1, R, $[M^2]^+$, $-M^2X$, inversion, R, L_nM^1

L_nM^1, R, $[M^2]^+$, +EX, $-M^2X$, *cis*, E, R, L_nM^1

Scheme 1

2. 1,2-Metallate Rearrangements of α-Alkoxyalkenyl Borates

In 1976 Levy and Schwartz [4] suggested that the borate **2** generated at -80°C by reaction of α-methoxyvinyl lithium (**1**) and tri-isobutylborane underwent a 1,2-metallate rearrangement on warming to room temperature to produce the alkenyl borate **3** (Scheme 2) whose structure was inferred from the formation of ketone **4** on standard oxidative workup. A far more interesting transformation resulted from treatment of the putative intermediate **3** with MeI prior to oxidative workup. In this case, the tertiary alcohol **6** was isolated in 89% yield and its formation demanded an unprecedented second 1,2-metallate rearrangement (**3** to **5**) of uncommon ease. However, further experiments in analogous systems showed that α-alkoxyalkenyl borates are far more stable than previously supposed and that 1,2-metallate rearrangements can take place by two different mechanisms depending on the reaction conditions [5][6].

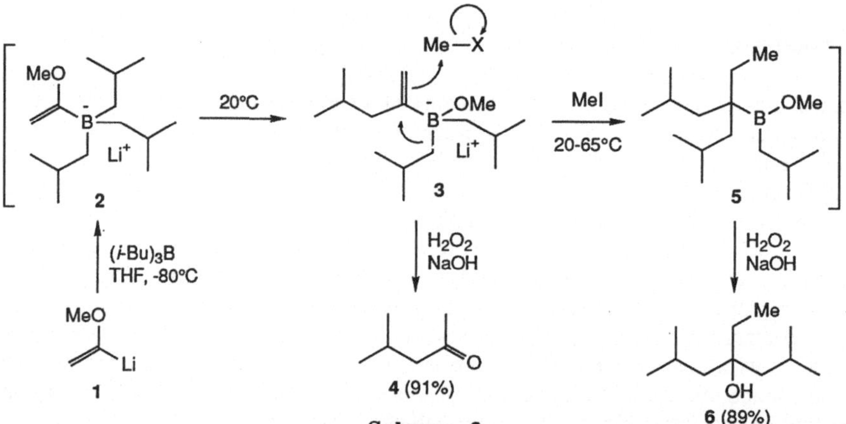

MeO
B
Li⁺
2
20°C
Me—X
OMe
B
Li⁺
3
MeI
20-65°C
Me
OMe
B
5
$(i\text{-Bu})_3B$
THF, -80°C
MeO
Li
1
H_2O_2
NaOH
O
4 (91%)
H_2O_2
NaOH
Me
OH
6 (89%)

Scheme 2

Addition of various trialkylboranes to a solution of 5-lithio-2,3-dihydrofuran **7** gave the borates **8a-d** which were then warmed to 20°C whereupon oxidative workup (Scheme 3) gave the hydroxyketones **10a-d**. Similarly, heating the borates **8a** with methyl iodide followed by HOAc and oxidative workup provided the tertiary alcohol **13**. We presumed that the borates **8a-d** rearranged on warming to give the alkenylborates **9** in accord with Levy's postulate and that these then underwent the observed transformations. However, all attempts to perform other reactions characteristic of alkenyl boranes (*vide infra*) using the putative intermediate **9** failed completely. These experiments suggested that the α-alkoxyalkenyl borates **8a-d** were stable and that rearrangement to **9** was not taking place under the reaction conditions. Indeed, the borates **8a-d** could be heated at 65°C (THF) for 24 h without suffering any further change. Consequently the 1,2-alkyl shifts must have taken place by electrophilic attack on the borates **8a-d**; furthermore, the fact that rearrangement was induced by aqueous NaOH and MeI suggests that the borates **8a-d** were very susceptible to electrophilic attack at the alkene terminus.

The formation of the methylated tertiary alcohol **13** (Scheme 4) required a second 1,2-alkyl shift induced by reaction of intermediate **14** with HOAc. In order to prove that heating borates with alkyl halide was sufficient to induce a single 1,2-alkyl migration, we treated the α-alkoxyalkenyl borate **16** with MeI and allyl bromide at 20°C for 16 h to induce alkylative rearrangement and then oxidised the intermediate **17** to give the hydroxyketones **18a,b** in modest yield.

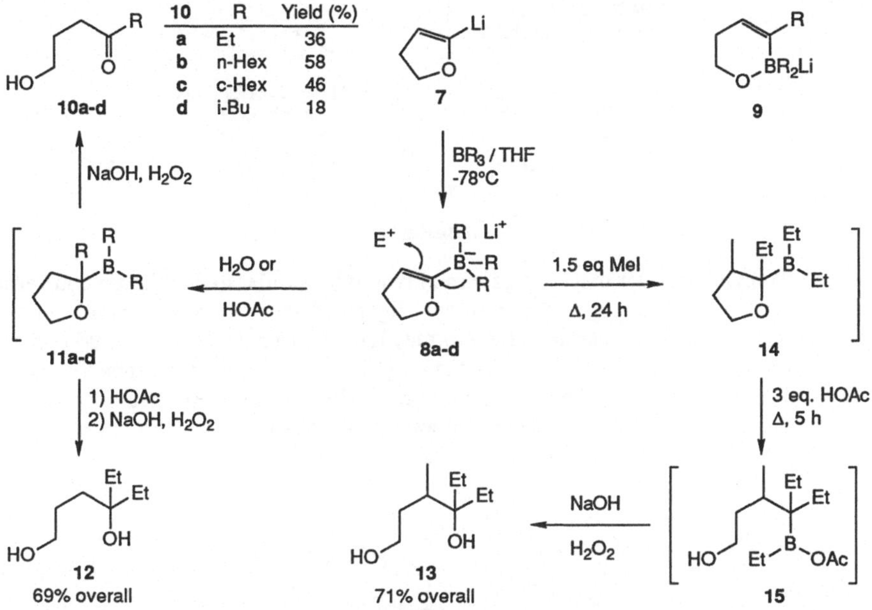

10	R	Yield (%)
a	Et	36
b	n-Hex	58
c	c-Hex	46
d	i-Bu	18

Scheme 3

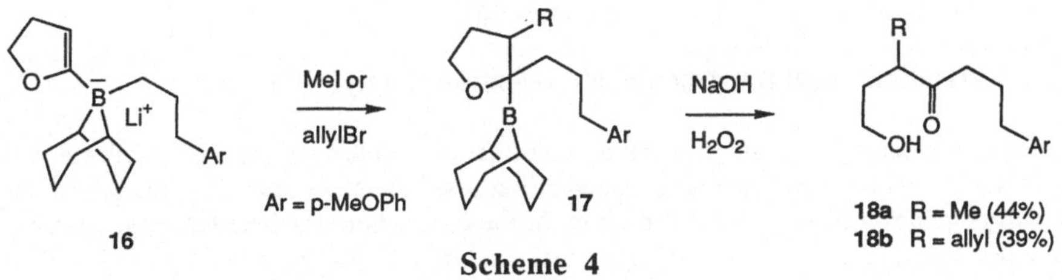

Scheme 4

Scheme 5 illustrates an alternative mechanism by which a 1,2-alkyl shift can be induced. When α-alkoxyalkenyl borate **8b** was heated for 24 h in THF in the presence of 1 equivalent of the oxyphilic Lewis acid TMSCl [5], an intermediate **19** was generated which displayed chemistry typical of alkenylboranes [7]. Thus treatment of **19** with iodine followed by oxidative workup generated the trisubstituted alkene **20**. In addition, the "ate" complex derived from reacting **19** with one equivalent of *n*-BuLi, transmetallated to an organocuprate which then coupled [8] with MeI to give the alkene **21** in 38% yield. Similarly, Suzuki coupling [9] converted **19** to the diene **22** in 55% yield. In both coupling reactions, a single stereoisomer was isolated. Unfortunately attempts to apply the reactions outlined in Scheme 5 to the borates derived from the corresponding dihydropyrans failed.

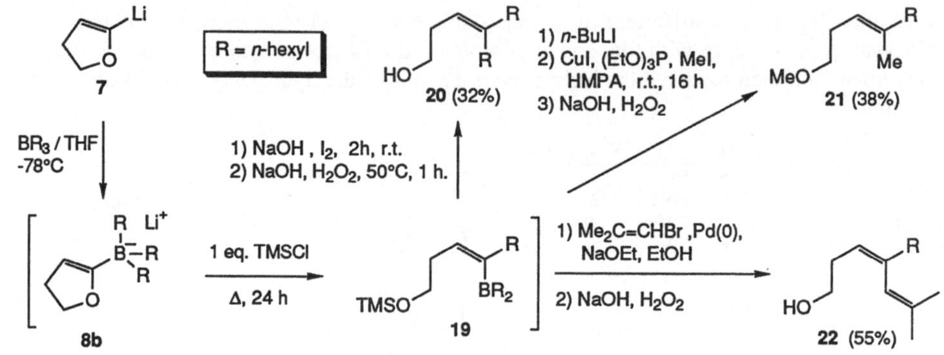

Scheme 5

α-Alkoxyalkenylaluminates display a pattern of reactivity similar to their boron counterparts. Thus the aluminate **23** bearing a dihydrofuranyl ligand (Scheme 6) was unstable and underwent an easy 1,2-metallate rearrangement to the alkoxyaluminate **24** under mild conditions and subsequent protonolysis gave the (*E*)-alkenol [10]. However, the corresponding dihydropyranyl substituted aluminate **25** was stable until challenged by a powerful electrophile in which case a 1,2-alkyl shift was induced without displacement of the adjacent C-O bond as exemplified by the formation of **26**.

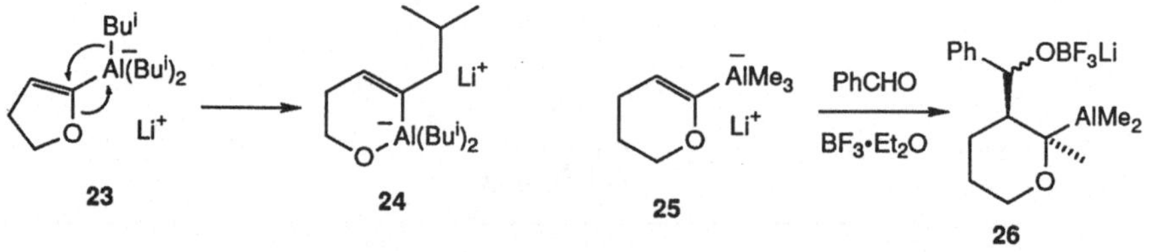

Scheme 6

3. 1,2-Metallate Rearrangements of α-Aminoalkenyl Borates

The foregoing experiments suggest that a functional ensemble comprised of an alkene bearing the combination of a metallate complex and a first row heteroatom bound to the α-carbon displays high electron density at the β-carbon. In the case where the heteroatom is oxygen, we have seen nucleophilic behaviour reminiscent of enamines – as first noted by Levy – in which even relatively weak electrophiles such as lithium cation or MeI are sufficient to induce a 1,2-alkyl shift. If the donation of the a non-bonded electron pair from the heteroatom to the alkene is a prime contributor to the nucleophilicity of the ensemble, we would expect that the corresponding nitrogen analogues would be even more reactive reflecting the intrinsically higher nucleophilicity of enamines compared with enol ethers. The evidence available to date in support of such a supposition comes from the very efficient and easy rearrangements of trialkyl (1-methyl-2-indolyl)borates [11]. For example, the borate **27** reacts with a wide range of electrophiles including primary alkyl triflates [12] to give 2,3-disubstituted indoles after oxidative workup. The reaction involves an electrophile-induced 1,2-metallate rearrangement leading to the creation of two new C-C bonds at C-2 and C-3 as illustrated in Scheme 7.

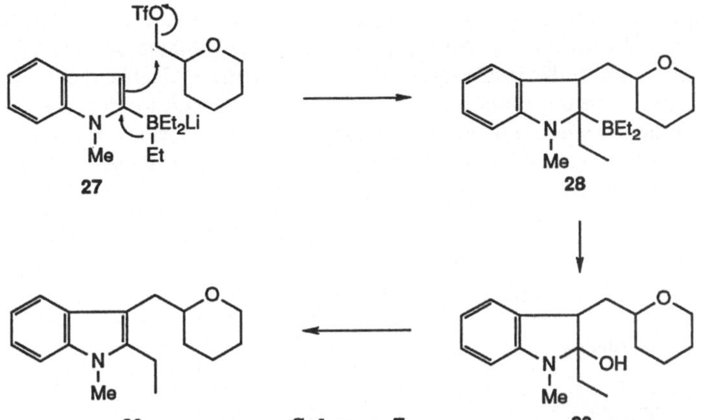

Scheme 7

A striking demonstration of the high nucleophilicity of α-aminoalkenylborates is shown in Scheme 8 in which a one-pot cyclisation procedure for [b]-annulated indole **34** was developed involving intramolecular reaction of a trialkyl-(1-methyl-2-indolyl)borate (where R represents the 9-BBN skeleton) with a π-allylpalladium complex **32** generated *in situ* [13].

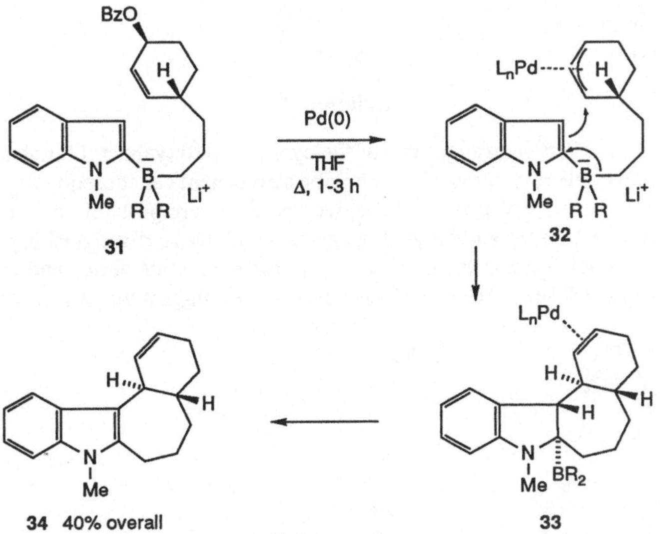

Scheme 8

4. 1,2-Metallate Rearrangements of α-(Carbamoyloxy)alkenyl Borates

All of the 1,2-metallate rearrangements we have cited thus far have involved *cyclic* α-heteroalkenyl borates and it was of interest to determine whether or not the reaction could be extended to acyclic systems which lack the stereochemical constraints imposed by the ring. Although progress has been hampered by the paucity of methods for preparing acyclic metallated enol ethers, we have briefly examined the chemistry of the α-(carbamoyloxy)alkenyl borate **36** (Scheme 9) which was readily prepared from the lithiated enol carbamate **35** [14] [15].Unlike the cyclic α-alkoxyalkenyl borates (see Scheme 3), HOAc did not provoke rearrangement of **36**; instead protonolysis of a C-B bond occurred to give the borane **37** as a stable entity isolated by column chromatography. However, rearrangement did occur *in the absence of TMSCl* on heating **36** to 50°C for 3 h whereupon the alkenyl borane **38** was generated which displayed typical reactions (Scheme 9).

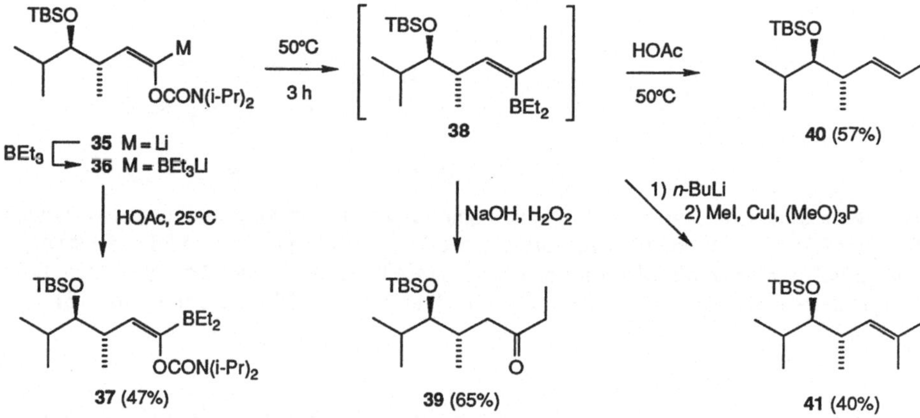

TBSO M OCON(i-Pr)$_2$ 50°C 3 h TBSO BEt_2 **38** HOAc 50°C TBSO **40** (57%)

BEt_3 **35** M = Li **36** M = BEt_3Li

HOAc, 25°C NaOH, H_2O_2 1) *n*-BuLi 2) MeI, CuI, $(MeO)_3P$

TBSO BEt_2 OCON(i-Pr)$_2$ **37** (47%) TBSO O **39** (65%) TBSO **41** (40%)

Scheme 9

Although the 1,2-metallate rearrangments of the cyclic α-alkoxyalkenyl borates and the acyclic α-(carbamoyloxy)alkenyl borates had the same stereochemical consequence–displacement with clean inversion–the mechanisms of the two processes are probably different. The easy electrophilic attack by protons and alkylating agents which was a distinguishing feature in the cyclic α-alkoxyalkenyl series was not observed in the carbamate series and therefore rearrangement by path A (Scheme 10) is unlikely. Instead, we suggest that the 1,2-metallate rear-

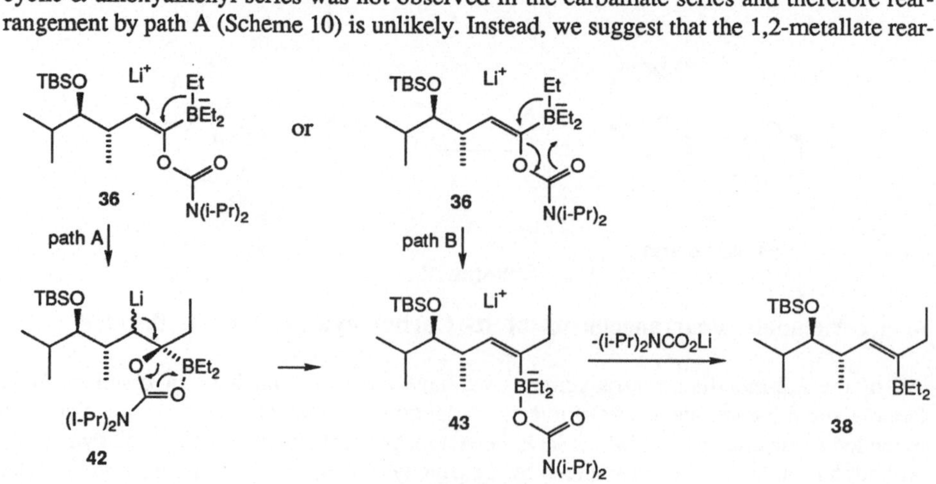

TBSO Li^+ Et BEt_2 O O **36** N(i-Pr)$_2$ or TBSO Li^+ Et BEt_2 O O **36** N(i-Pr)$_2$

path A path B

TBSO Li O BEt_2 O (I-Pr)$_2$N **42** → TBSO Li^+ BEt_2 O O **43** N(i-Pr)$_2$ -(i-Pr)$_2NCO_2Li$ TBSO BEt_2 **38**

Scheme 10

rangment is diverted through path B–a dyotropic rearrangement [16]–in order to accomodate both the diminished nucleophilicity of the alkene moiety and the superior nucleofugacity of the carbamate moiety.

5. 1,2-Metallate Rearrangements of Cyclic α-Alkoxyalkenyl Lithio Cuprates

It has been known for more than forty years that α-unsubstituted cyclic enol ethers undergo reaction with a variety of organometallics. For instance, in 1952 Paul and Tchelitcheff [17] reported that reaction of 3,4-dihydro-2H-pyran **42** with pentylsodium gave rise to the relatively stable α-metallated species **43** which could be trapped with carbon dioxide to give carboxylate **44** (Scheme 11). Similarly, metallation/carboxylation of 2,3-dihydrofuran occurred exclusively at the α–position. With less basic reagents such as Grignard reagents or alkyllithium reagents a different pattern of reactivity emerged. Hill and co-workers [18] found that prolonged exposure of **42** to *n*-octylmagnesium bromide gave coupling product **45** in modest yield, and Pattison and Dear [19] reported that the same transformation occurred more efficiently (73%) using *n*-BuLi. Furthermore, the coupling product **46** was isolated as single (*E*)-isomer. Schlosser [20] showed that *t*-BuLi could be used to similar effect.

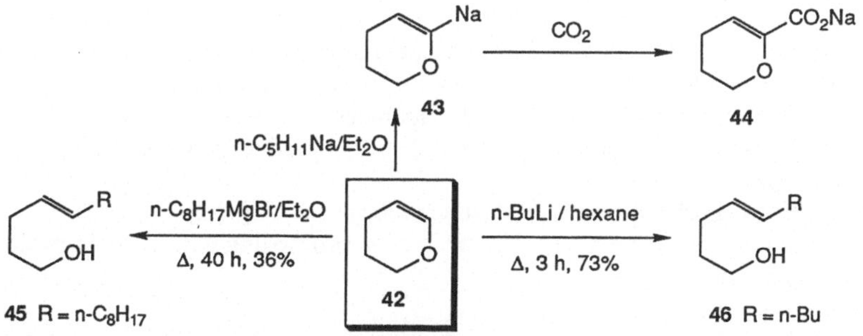

Na CO$_2$ CO$_2$Na
O O
43 44
n-C$_5$H$_{11}$Na/Et$_2$O
R n-C$_8$H$_{17}$MgBr/Et$_2$O n-BuLi / hexane R
OH Δ, 40 h, 36% O Δ, 3 h, 73% OH
45 R = n-C$_8$H$_{17}$ 42 46 R = n-Bu

Scheme 11

Even more intriguing was the discovery, made by Fujisawa and his collaborators [21], that the reaction of 2,3-dihydrofuran **47** with alkyl-lithiums was greatly facilitated by addition of one equivalent of CuI; the coupling proceeded with very high stereoselectivity and under comparatively mild conditions to give the (*E*)-alkenol in excellent yield (Scheme 12). In each of the foregoing transformations an apparent nucleophilic displacement of an alkyl-ether bond has taken place with inversion of stereochemistry. Surprisingly, Pattison and Dear were the only investigators to proffer a mechanistic rationale for these connective ring cleavage reactions of cyclic enol ethers until 1988 when Sjoerd Wadman performed one simple experiment which illuminated the mechanism of Fujisawa's reaction and its forebears and ultimately led to the discovery of a novel 1,2-metallate rearrangement of α-alkoxyalkenyl lithio cuprates of considerable synthetic potential [22]. Firstly, Wadman showed (Scheme 12) that quenching the reaction mixture with D_2O led to stereo- and regiospecific introduction of a deuterium atom onto the alkene and secondly he showed that 5-lithio-2,3-dihydrofuran (**7**) reacted with 2 equivalents of *n*-Bu_2CuLi to give the deuterated product **48** after quenching with D_2O.

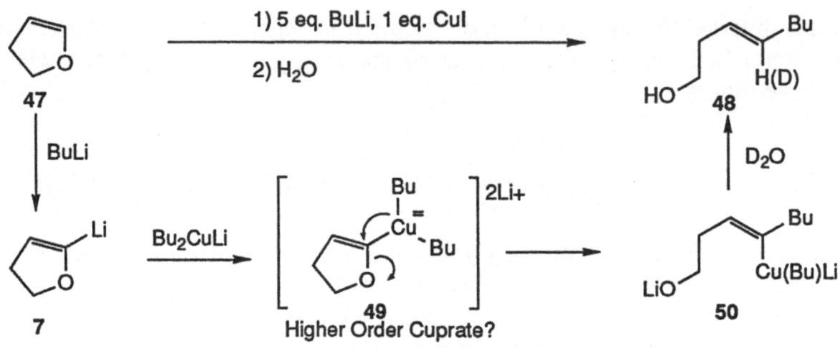

1) 5 eq. BuLi, 1 eq. CuI

2) H_2O

47

BuLi

7

Bu_2CuLi

49
Higher Order Cuprate?

2Li+

48

H(D)

D_2O

50

Cu(Bu)Li

Scheme 12

What is the relation between the earlier experiments of Hill, Pattison, and their co-workers and the reaction described by Fujisawa? To establish whether or not the presence of Cu(I) salts caused an obvious change in mechanism, we repeated the reactions reported by Hill [18] and Pattison [19], quenching with deuterium oxide. Once again, coupling products **45** and **46** were fully deuterated after both experiments - suggesting that the mechanisms of all the coupling reactions were similar, regardless of the metal used. We can summarise the observations discussed in the foregoing section as follows:

(i) Coupling of 3,4-dihydro-2H-pyran (**42**) and 2,3-dihydrofuran (**47**) occurred with alkyl-lithium and alkylmagnesium derivatives *in the absence of* Cu (I) salts.
(ii) The presence of Cu(I) salts greatly facilitated the reaction in all cases.
(iii) The α-metallated cyclic enol ethers (e.g. **7**) were intermediates on the reaction pathway.
(iv) The final products were alkenylmetallic species (e.g. **50**).
(v) The reactions were highly stereoselective proceeding with clean inversion of configuration.

A number of possible mechanisms leading to **46** can be envisaged, including ones which bear a resemblance to the boron-mediated rearrangements described in section 2, but our results do not enable us to prove or disprove any of these. The mechanism which, in our view, accounts most comfortably for all our observations is outlined in Scheme 12. The key aspect of this hypothesis is that rearrangement occurs within the coordination sphere of a mixed metal "ate" complex **49**, by a nucleophilic displacement at an sp^2 center. The 1,2-metallate rearrangement is depicted in Scheme 12 as being more or less concerted, but a stepwise sequence involving a vinylidene metal carbenoid cannot be discounted (*vide infra*). The virtue of Scheme 12 is that it readily accounts for the observed stereoselectivity without making further assumptions necessary. However, there are two objections which can be raised against the mechanism: first, we implicate a higher order cuprate intermediate (**49**) whose existence has previously been challenged [23][24] and secondly, direct displacement at sp^2 centres are very rare [25].

In order to facilitate further discussion of the scope, mechanism, and synthetic applications of 1,2-metallate rearrangements of organocuprates, we will now define a 'standard' 6-step-sequence for the connective synthesis of functionalised trisubstituted alkenes (Scheme 13) which will be followed later by some useful variations. The sequence begins (step A, Scheme

13) with the conversion of an alkyl bromide or iodide to the corresponding organolithium reagent which is easily achieved by treatment of the alkyl halide with 2 equivalents of *t*-BuLi in THF at -78°C [26]. The solvent is then removed *in vacuo* and replaced by dry Et_2O. The resultant solution is then added dropwise to a slurry of CuCN in Et_2O at -30°C and then, after 20 min, the cooling bath is removed to form the lower order cyanocuprate **51** (step B). In the meantime, in a separate vessel, the enol ether **52** is metallated (step C) by adding 1 equivalent of *t*-BuLi/pentane to a solution of the enol ether in THF at -30°C [26]. On warming to 0°C, the yellow-orange colour of the *t*-BuLi•2THF complex fades giving a colourless or pale yellow solution of the lithiated enol ether **53** from which the solvent is removed *in vacuo* and replaced by Et_2O. The solution of the lithiated enol ether **53** is then added *rapidly* to the lower order cuprate to form the higher order cuprate intermediate **54** (step D) [28] and the resultant mixture refluxed for 2 h during which time the 1,2-metallate rearrangement takes place (step E) to give the alkenylcuprate **55**. To complete the sequence, intermediate **55** can be quenched with a suitable electrophile (step F) to give the trisubstituted alkene **56**.

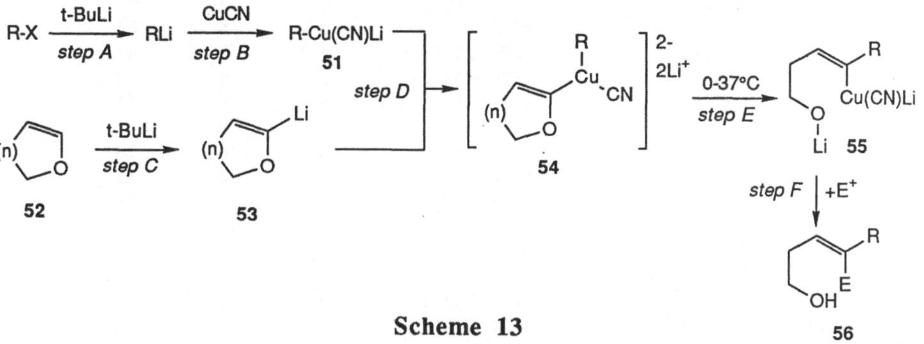

R-X —(t-BuLi, step A)→ RLi —(CuCN, step B)→ R-Cu(CN)Li (**51**)

(n)-ring enol ether **52** —(t-BuLi, step C)→ lithiated enol ether **53**

51 + **53** —(step D)→ [cuprate **54**, R–Cu–CN]$^{2-}$ $2Li^+$ —(0-37°C, step E)→ **55** (Cu(CN)Li, OLi) —(step F, $+E^+$)→ **56** (R, E, OH)

Scheme 13

Convenience or necessity may demand variations on the "standard" procedure described above. The most significant variations to the protocol are dictated by the need to achieve the optimum temperature for the 1,2-metallate rearrangement (step E). If the temperature is too low (e.g., -30°C) rearrangement does not take place; however, if the temperature is too high, decomposition of either **51** or **54** may lead to a multitude of minor products. The optimum temperature depends on the Cu(I) salt used, the solvent, the structure of the alkyl ligand (*t*-Bu reacts faster than *n*-Bu) and the structure of the enol ether (dihydrofurans react at 0°C whereas dihydropyrans may require room temperature or even 37°C). Table 1 illustrates the influence of temperature (entries 1-4), stoichiometry (entries 5-7) and catalyst (entries 5, 8, and 9) on the yield of the rearrangement product **58a** derived from reaction of 6-lithio-3,4-dihydro-2H-pyran (**57**) with *n*-BuLi. Although temperature had a dramatic effect, the Cu(I) catalyst would appear to be comparatively insignificant when simple alkyl-lithiums are used. However, in some cases reproducibility is enhanced by switching from CuCN/Et_2O to CuBr•SMe_2/Et_2O-SMe_2 because in the latter case, the cuprates are more often homogeneous. Moreover, the stabilising effect of SMe_2 on organocuprates has been previously noted [29]. Interestingly the soluble CuCN•2LiCl complex [30] gave consistently poor results.

Two further generalisations can be gleaned from the results summarised in Tables 1 and 2. First, the scope of the reaction is quite broad: primary, secondary, tertiary alkyl, cyclopropyl,

Table 1. Cu(I)-Catalyzed Reaction of Organolithium Reagents with 6-Lithio-3,4-dihydro-2H-pyran (57)

57 → 1) RLi, CuX / Et_2O; 2) H_2O → 58a-k

Entry	R(Li)(Equiv)	CuX (Equiv)	Temperature (°C)	58	Yield (%)
1	n-Bu (2.0)	CuCN (1.0)	-30	a	0
2	n-Bu (2.0)	CuCN (1.0)	-10	a	27
3	n-Bu (2.0)	CuCN (1.0)	0	a	45
4	n-Bu (2.0)	CuCN (1.0)	20	a	61
5	n-Bu (1.0)	CuCN (0.5)	Δ	a	58
6	n-Bu (2.0)	CuCN (0.5)	Δ	a	68
7	n-Bu (4.0)	CuCN (1.0)	Δ	a	78
8	n-Bu (1.0)	$CuBr{\bullet}SMe_2$ (0.5)	Δ	a	52
9	n-Bu (1.0)	CuI (0.5)	Δ	a	63
10	s-Bu (1.0)	CuCN (0.5)	Δ	b	47
11	s-Bu (2.0)	CuCN (0.5)	Δ	b	70
12	t-Bu (1.0)	CuCN (0.5)	Δ	c	64
13	t-Bu (2.0)	CuCN (0.5)	Δ	c	77
14	Me (2.0)	CuCN (0.5)	Δ	d	10
15	Ph (2.0)	CuCN (0.5)	Δ	e	15
16	CH_2=C(Me) (4.0)	CuCN (0.07)	Δ	f	79
17	$PhMe_2Si$ (4.0)	CuCN (0.2)	Δ	g	92
18	Me_3Sn (4.0)	CuCN (0.2)	Δ	h	78
19	c-C_3H_5 (1.0)	CuCN (0.5)	Δ	i	33
20	o-MeOPh (1.0)	CuCN (0.5)	Δ	j	0
21	p-MeOPh (1.0)	CuCN (0.5)	Δ	k	33

Table 2. Cu(I)-Catalyzed Reaction of Organolithium Reagents with 5-Lithio-2,3-dihydrofuran (7)

7 → 1) RLi, CuX / Et_2O; 2) H_2O → 48a-e

Entry	R(Li)(Equiv)	CuX (Equiv)	Temperature (°C)	48	Yield (%)
1	n-Bu (1.0)	CuCN (0.5)	0	a	58
2	n-Bu (4.0)	CuCN (0.1)	0	a	82
3	s-Bu (1.0)	CuCN (0.5)	0	b	60
4	s-Bu (4.0)	CuCN (0.1)	0	b	88
5	t-Bu (1.0)	CuCN (0.5)	0	c	77
6	t-Bu (4.0)	CuCN (0.1)	0	c	77
7	Me (1.0)	CuCN (0.5)	0	d	7
8	o-MeOPh (1.0)	CuCN (0.5)	0	e	47

aryl, and vinyl lithiums participate in the coupling reaction as do phenyldimethylsilyl- and trimethylstannyl-lithium. Of the common lithium reagents examined Me and Ph were exceptional in giving low yields (Table 1, entries 14 and 15; Table 2, entry 7); allyl-lithium gave a very messy reaction, and homoallylic lithium reagents refused to participate. Secondly, the reaction is catalytic in Cu(I) with as little as 0.1 equiv. of the catalyst being required for satisfactory results. Hence, a complex series of transmetallation reactions must be involved as summarised in the catalytic cycle shown in Scheme 14. Thus, by simply varying the stoichiometry of the process, either an alkenyl-lithium **60** or an alkenyl-cuprate **59** can be generated at will.

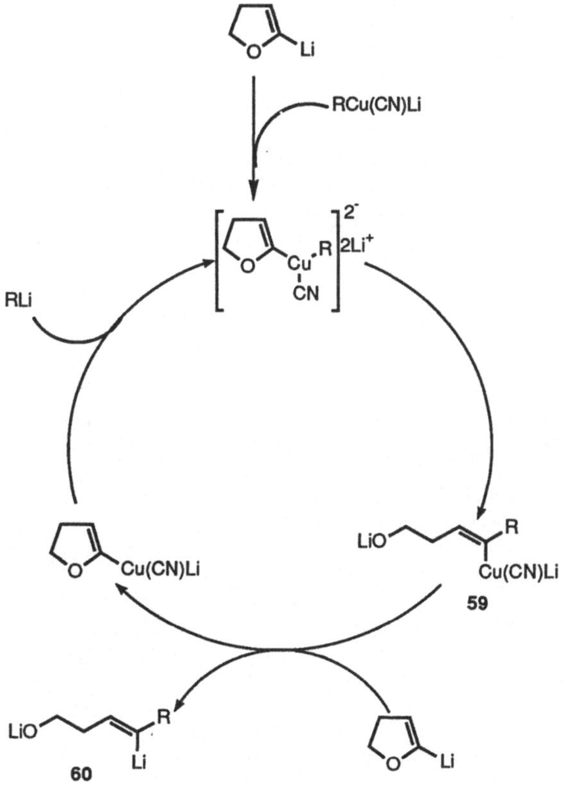

Scheme 14

Our original aim in developing the reactions described herein was the stereoselective synthesis of trisubstituted alkenes in which case success would ultimately depend upon an efficient alkylation of the alkenylmetallic species (step F, Scheme 13). We have invested comparatively little effort in this aspect of the sequence but preliminary results indicate that the alkenylmetallic reagents are sluggish in their reactions with electrophiles. Nevertheless, alkylation can be achieved with reactive electrophiles such as MeI in the presence of HMPA as illustrated by the synthesis of the polyketide fragment of the potent immunosuppressant FK-506 (Scheme 15) [31].

In all of the foregoing reactions we have employed 5-lithio-2,3-dihydrofuran (7) and 6-lithio-3,4-dihydro-2H-pyran (**57**) as the enol ether partner. It was of interest to explore alternative α-heteroalkenyl-lithium reagents but results so far have been disappointing. Listed below are some reagents which fail to participate in the coupling and it is noteworthy that it takes little by way of oxygen substitution to preclude reaction in dihydropyrans **61** and **62**. Furthermore, for reasons which have yet to be discovered, none of the acyclic enol ethers studied thus far (**64a**, **65**, **66**) have undergone coupling.

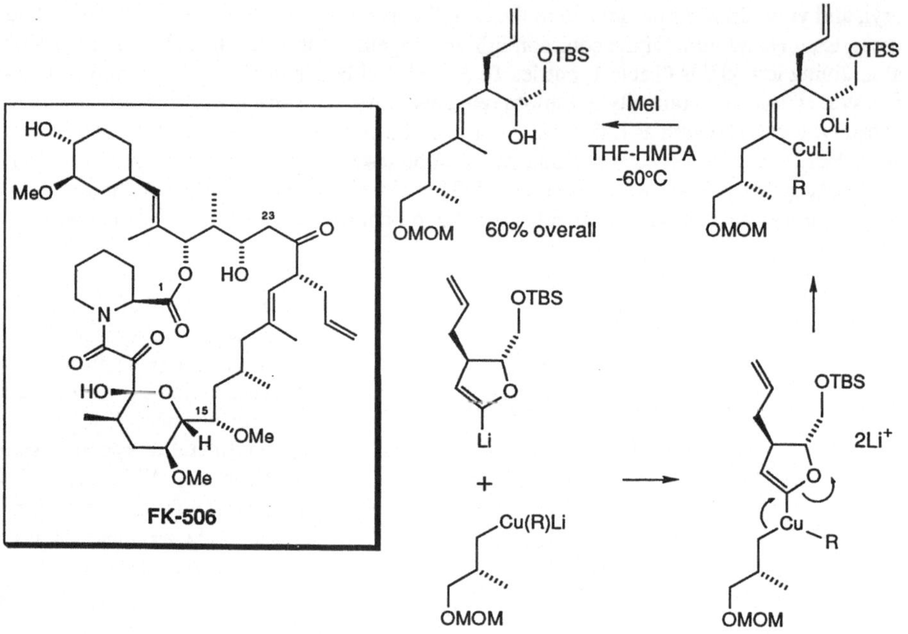

Scheme 15

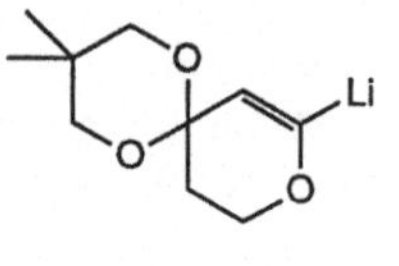

61 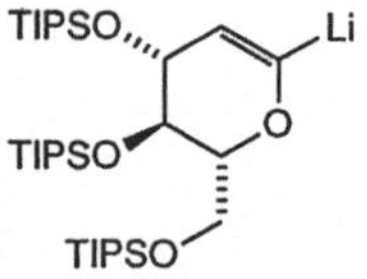 62 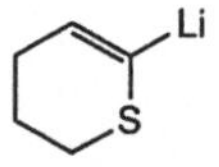 63 64 65 66

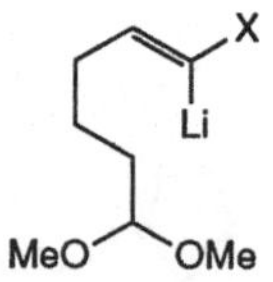

64 a X = OMe b X = SPh c X = SOPh d X = SO_2Ph

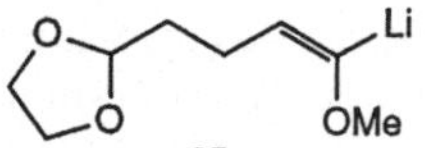

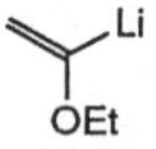

6. 1,2-Metallate Rearrangements of α-Alkoxyalkenyl Magnesio Cuprates.

The CuCN-catalysed coupling of organolithium reagents with 5-lithio-2,3-dihydrofuran and 6-lithio-3,4-dihydro-2H-pyran is subject to a number of limitations. We have already seen that four equivalents of cyanocuprate were required for optimum yields leading to conspicuous waste in the case of expensive ligands. We have also seen that allylic and homoallylic cuprates are amongst the few reagents which fail to undergo coupling presumably because the requisite 1,2-metallate rearrangement required temperatures at or above room temperature in which case competing destruction of the cuprate was observed leading to low yields and messy reactions. Another problem which cannot be ignored is the comparative inaccessibility of many lithium reagents. In certain cases all these limitations can be surmounted by the use of Grignard reagents in place of organolithium reagents in the preparation of the organocuprate precursor [32].

The practical advantages of using Grignard reagents as reaction partners in the Cu(I)-catalysed coupling with α-lithiated enol ethers is illustrated by the transformation shown in Scheme 16 - a transformation which we tried in vain to accomplish according to our original protocol using cuprates derived from organolithiums. The homoallylic Grignard **67** reagent was added to a suspension of CuCN in THF. To the resultant cyanocuprate **68** was added a solution of **7** at -80°C and the mixture allowed to warm slowly to 0°C during which time the 1,2-metallate rearrangement took place to produce the alkenylcuprate **69** which was quenched with the indicated electrophiles to give the homoallylic alcohols **70-72** in 75% yield.

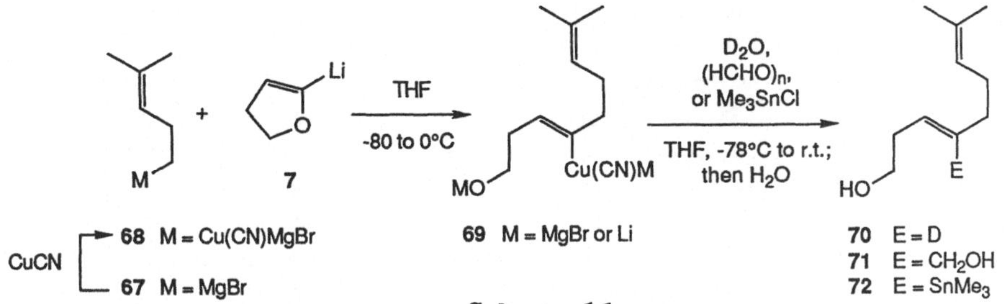

THF, -80 to 0°C; D_2O, $(HCHO)_n$, or Me_3SnCl; THF, -78°C to r.t.; then H_2O

CuCN: **68** M = Cu(CN)MgBr; **67** M = MgBr; **7**; **69** M = MgBr or Li; **70** E = D; **71** E = CH_2OH; **72** E = $SnMe_3$

Scheme 16

In order to establish the scope and limitations of the reaction depicted in Scheme 16 we examined similar transformations with a variety of Grignard reagents and 5-lithio-2,3-dihydrofuran (**7**) or 6-lithio-3,4-dihydro-2H-pyran (**57**) in Et_2O using CuBr•SMe_2 complex as the source of Cu(I) to give the corresponding substituted (*E*)-3-buten-1-ols **48a-c,f-h** and (*E*)-4-penten-1-ols (**58a-f,l-q**) respectively. We optimised the reaction for 1 equivalent each of the Grignard reagent and the α-metallated enol ether even though superior yields could be obtained by using an excess of Grignard reagent (compare entries 4 and 5, Table 4). The following general observations are pertinent to the results which are summarised in Tables 3 and 4.

(i) In every case examined, the *trans*-1,2-disubstituted alkenes **48** and **58** were formed with ≥97% stereoselectivity according to high field ^1H and ^{13}C NMR spectroscopy.

(ii) Fewer side reactions and greater reproducibility were obtained when CuBr•SMe_2 complex was used instead of CuCN probably because of the greater solubility of the former. Although as little as 10 mol% of Cu(I) could be used with only a negligible loss in yield, the reactions were cleaner–and therefore product purification easier–when 0.5 equiv was used with **57** or 1.0 equiv with **7**.

(iii) The efficiency and rate of the 1,2-metallate rearrangements depended on the structure of the metallated enol ether as well as the structure of the organometallic ligand. Fastest rates, cleanest reactions, and best yields were obtained with ter-

Table 3. Cu(I)-catalysed Reaction of Grignard Reagents with 6-Lithio-3,4-dihydro-2H-pyran (**57**).

1) RMgX (1 equiv)
$CuBr\cdot Me_2S$ (0.5 equiv)
Et_2O, -40 to -30°C, 2-3 h.
2) H_2O

57 (1 equiv) → **58a-f,l-q**

Entry	R	X	**58**	Yield(%)[a]
1	n-Bu	Br	**a**	48
2	s-Bu	Cl	**b**	58
3	t-Bu	Cl	**c**	64
4	Me	Br	**d**	29
5	Ph	Br	**e**	46
6	CH_2=C(Me)	Br	**f**	32
7	Et	Br	**l**	40
8	n-Hex	Br	**m**	64[b]
9	c-Hex	Cl	**n**	54
10	i-Pr	Cl	**o**	60
11	allyl	Br	**p**	44
12	2-methylallyl	Cl	**q**	46

[a] Yields refer to products purified by column chromatography followed by distillation.
[b] Grignard reagent : **10** = 3 : 2

Table 4. Cu(I)-catalysed Reaction of Grignard Reagents with 5-Lithio-2,3-dihydrofuran (**7**).

1) RMgX (1 equiv)
$CuBr\cdot Me_2S$ (1 equiv)
Et_2O, -70°C, 1-2 h.
2) H_2O

7 (1 equiv) → **48a-c,f-h**

Entry	R	X	**48**	Yield(%)[a]
1	n-Bu	Br	**a**	31
2	s-Bu	Cl	**b**	33
3	t-Bu	Cl	**c**	33
4	n-Hex	Br	**f**	38
5	n-Hex	Br	**f**	49[b]
6	c-Hex	Cl	**g**	84
7	Ph	Br	**h**	50

[a] Yields refer to products purified by column chromatography followed by distillation.
[b] Grignard reagent : **3** = 3 : 2

tiary and secondary alkyl Grignard reagents. In contrast to the lithiocuprates reported previously, the magnesiocuprates gave better yields with the dihydropyran **57** than the dihydrofuran **7**.

(iv) The higher order cuprates derived from organolithiums, lithiated enol ethers **57** or **7**, and $CuBr{\bullet}SMe_2$ rearranged at appreciable rates at 0-35°C whereas the corresponding reactions with organomagnesium reagents occurred at -70°C in the case of **7** or *ca.* -30°C in the case of **57**. The significance of the metal cation in the course of the rearrangement was demonstrable. Thus, the reaction of **57** with *n*-$Bu_2CuLi{\bullet}SMe_2$ occurred normally at *ca.* -5°C and the rate was unaffected by addition of $MgBr_2{\bullet}Et_2O$. However, addition of 1 equivalent of LiCl to the reaction mixture derived from *s*-$Bu_2CuMgBr$ and **57** appreciably depressed the rate: the reaction now required a temperature of -5°C instead of -30°C – the temperature required for reaction in the absence of LiCl. Speculation on the origin of this marked effect is premature.

The principal limitation to the application of the Grignard variant of the 1,2-metallate rearrangement to the synthesis of trisubstituted alkenes is the sluggish rate at which the intermediate alkenylmetals (e.g. **69** in Scheme 16) undergo alkylation even with such reactive electrophiles as MeI.

7. 1,2-Metallate Rearrangements of α-(Carbamoyloxy)alkenyl Cuprates

In section 4 we noted that α-(carbamoyloxy)alkenyl borates undergo 1,2-metallate rearrangements with far greater ease than the corresponding reactions with α-alkoxyalkenyl borates. We speculated that the differences in reactivity between the two systems reflect the higher nucleofugacity of the carbamoyl group compared with an alkoxy group as well as a change in mechanism. We now demonstrate parallel behaviour in the 1,2-metallate rearrangements of α-(carbamoyloxy)alkenyl cuprates compared with α-alkoxyalkenyl cuprates and speculate that here too, there has been a change in mechanism.

Lithiation of Hoppe's (*Z*)-*anti* enol carbamate **73** [14] with *t*-BuLi at -85°C affords the lithiated species **35** in quantitative yield with complete retention of double bond configuration (Scheme 17). The lithio derivative was unstable and underwent slow Fritsch-Buttenberg-Wiechell rearrangement [33] even at -70°C; at -40°C the rearrangement to the alkyne **75** accelerated to an extent which thwarted attempts to prepare the higher order cyanocuprate **76** by reaction of **35** with *n*-BuCu(CN)Li. However, the corresponding alkenylstannane **74** prepared in 95% yield by reaction with Me_3SnCl at -85°C, transmetallated with 1.2 equivalent of *n*-$Bu_2Cu(CN)Li_2$ in Et_2O at 0°C to give the unstable higher order cyanocuprate intermediate **76** [34] which underwent a 1,2-metallate rearrangement. The resultant lower order cyanocuprate **77**, on protonolysis, afforded the homoallylic alcohol **78** (R = *n*-Bu) in 90% overall yield. Similar reactions of stannane **74** with the higher order cyanocuprates derived from *s*-BuLi, *t*-BuLi, and MeLi also gave good yields of the coupling product. Unfortunately the superior yields obtained in the carbamate-based rearrangements were accompanied by loss of stereoselectivity: with the exception of *t*-BuLi, all the coupling reactions gave mixtures of alkenes (*E:Z* = 4-5:1).

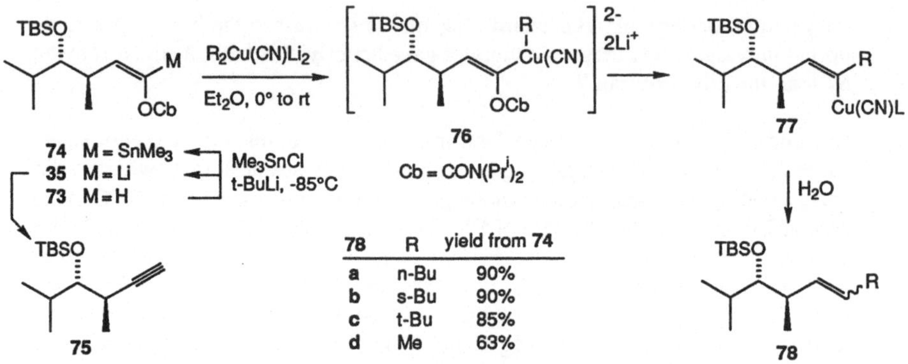

Cb = CON(Pr^i)$_2$

78	R	yield from 74
a	n-Bu	90%
b	s-Bu	90%
c	t-Bu	85%
d	Me	63%

Scheme 17

We believe that the diminished stereoselectivity in the formation of **78** and the attendant formation of alkyne **75** may unite both products by a common intermediate (Scheme 18) reflecting the intrinsic instability of α-(carbamoyloxy)alkenyl cuprates or their lithium precursor **83**. Thus α-elimination to the vinylidene carbene **80** provides a junction which leads either to alkenylcuprate product **82** by insertion into the C-Cu bond of a lower order cyanocuprate or to alkyne **81** by a Fritsch-Buttenberg-Wiechell rearrangement. Another possibility is that the alkenylcuprate **82** stems from a migratory insertion of vinylidene carbenoid **84** .

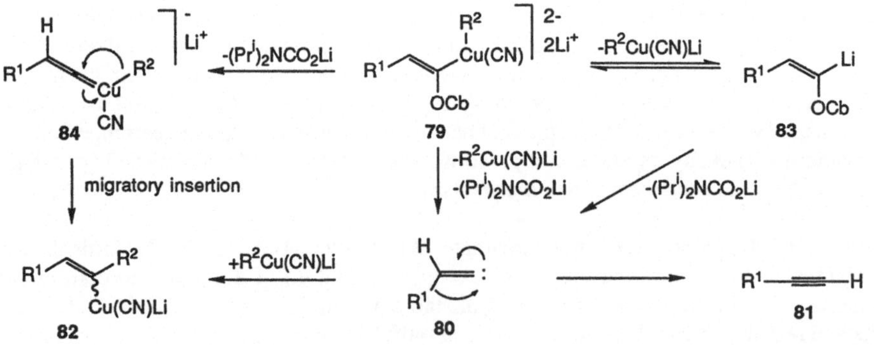

Scheme 18

Mr. Austen Pimm has recently exploited the 1,2-metallate rearrangement of α-(carbamoyloxy)alkenyl cuprates in a synthesis of the polyketide chain of the potent antifungal agent Jaspamide (Scheme 19). Starting with a homochiral (*Z*) *anti* enol carbamate prepared by Hoppe's homoaldol method, the requisite polyketide chain was synthesised in 11% overall yield in 12 steps.

OH
HN
HN
Br
MeN
NH
O
COOH
BzO

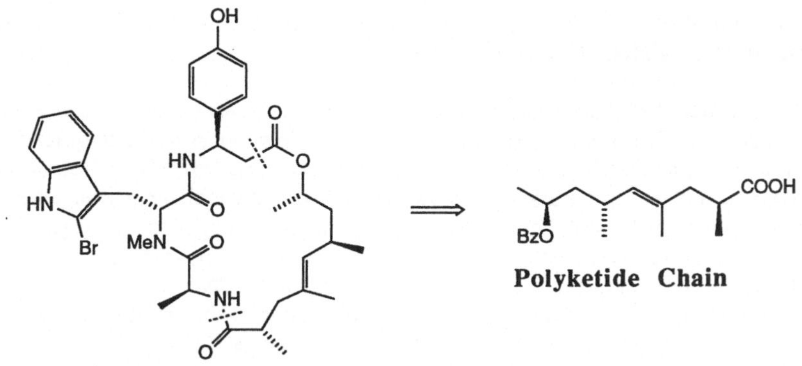

Jaspamide **Polyketide Chain**

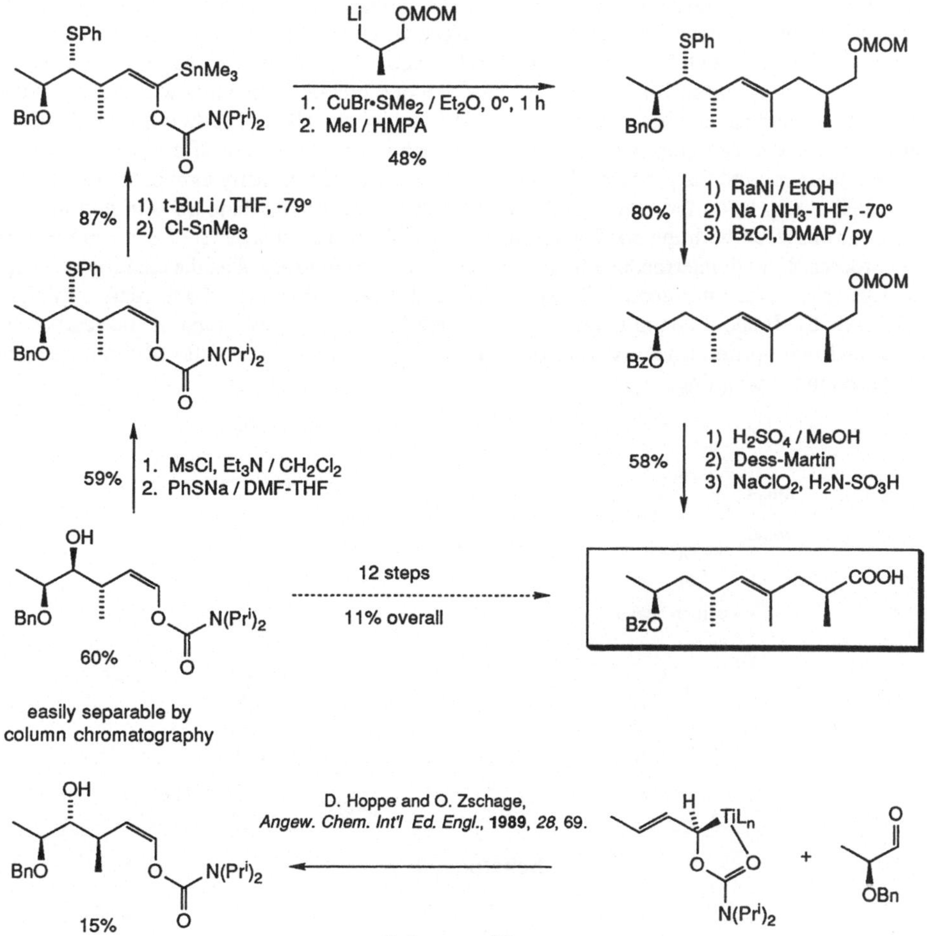

SPh
SnMe3
BnO
N(Pri)2
Li OMOM
1. CuBr•SMe2 / Et2O, 0°, 1 h
2. MeI / HMPA
48%
OMOM
87%
1) t-BuLi / THF, -79°
2) Cl-SnMe3
80%
1) RaNi / EtOH
2) Na / NH3-THF, -70°
3) BzCl, DMAP / py
BzO
59%
1. MsCl, Et3N / CH2Cl2
2. PhSNa / DMF-THF
58%
1) H2SO4 / MeOH
2) Dess-Martin
3) NaClO2, H2N-SO3H
12 steps
11% overall
COOH
60%
easily separable by
column chromatography
D. Hoppe and O. Zschage,
Angew. Chem. Int'l Ed. Engl., 1989, 28, 69.
TiLn
OBn
15%

Scheme 19

8. Do Cu(I)-Catalysed 1,2-Metallate Rearrangements of α-Alkoxyalkenyl Cuprates Proceed *via* Higher Order Cuprates?

In section 2 we showed that α-alkoxyalkenyl borates undergo 1,2-metallate rearrangement by two distinct pathways. Their thermal rearrangement is very slow unless the nucleofugacity of the alkoxy group is enhanced by silylation. On the other hand α-alkoxyalkenyl borates are highly nucleophilic and they are provoked into stereoselective 1,2-metallate rearrangement by very mild electrophiles such as MeI and water. The corresponding cuprates are far less stable and their 1,2-metallate rearrangements are easier. When we proposed that higher order cuprates are the crucial intermediates in the rearrangement when CuCN is used as the catalyst, we were mindful of Lipshutz' evidence [24] regarding the structure and stability of higher order cyanocuprates. However, CuBr and CuI are also effective catalysts and in some cases even superior to CuCN and yet the existence of higher order cuprates as stable entities has been questioned [23]. Nevertheless, there are two pieces of evidence which reconcile our proposal with new evidence regarding the stability of higher order cuprates. First, we noted that Me_2S frequently has a beneficial effect on the efficiency of the rearrangement; secondly, Olmstead and Power [35] have recently obtained single crystal x-ray structures of higher order cuprates *which incorporate Me_2S as stabilising ligands*. Using the x-ray data of Olmstead and Power we now propose that 1,2-metallate rearrangements of higher order cuprates proceed by either of two paths outlined in Scheme 20. Reaction of a lithiated enol ether with a lower order cuprate **85** generates the higher order cuprate **86** in which the alkenyl carbon is bound to copper as well as two bridging lithiums. A 1,2-alkyl shift with inversion of configuration may convert **86** directly to the product **87** – a transformation which is easily explicable in molecular orbital terms. If, on the other hand, the alkene bond of intermediate **86** is highly nucleophilic (as observed with the corresponding borates and aluminates), then the rearrangement may proceed through intermediate **88** in which a lithium cation served as the electrophilic trigger. Collapse of the intermediate **88** as shown would then account for the formation of alkenylcuprate **87**. It should be noted that intermediate **86** incorporates a total of four carbon ligands and that optimum yields were obtained when 4 equivalents of alkyl-lithium reagent were used in the coupling.

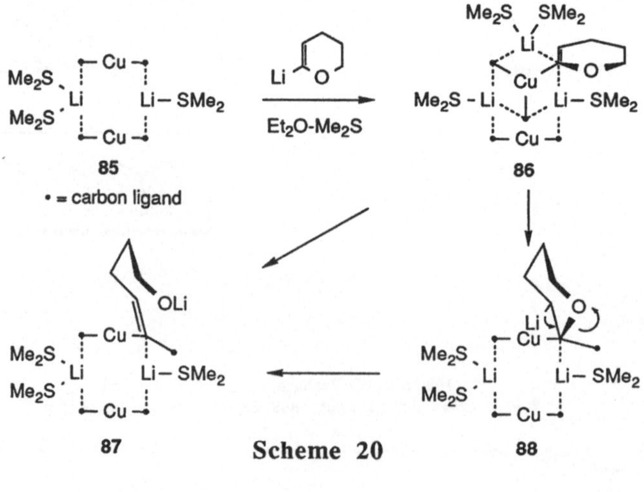

Me2S
Me2S
Li
Cu
Li–SMe2
Cu
85
• = carbon ligand
Li
O
Et2O-Me2S
Me2S SMe2
Li
Cu
O
Me2S - Li
Li– SMe2
Cu
86
OLi
Me2S
Me2S
Li
Cu
Li– SMe2
Cu
87
Li
O
Cu
Me2S
Me2S
Li
Li– SMe2
Cu
88

Scheme 20

9. Conclusion

At least four possible general mechanisms (paths A-D, Scheme 21) may be envisaged for the 1,2-metallate rearrangement of α-heteroalkenyl metallate complexes of the general structure **89** in which R = a migrating alkyl group; M = a metal capable of sustaining a negative charge (e.g. B, Al, Cu, Zn, Zr [36]); X = a heteroatom leaving group; L_n = ligand(s) attached to the metal; and E^+ = an electrophilic agent – be it a carbon electrophile such as MeI or a metal cation such as Li^+ or Mg^{2+}.

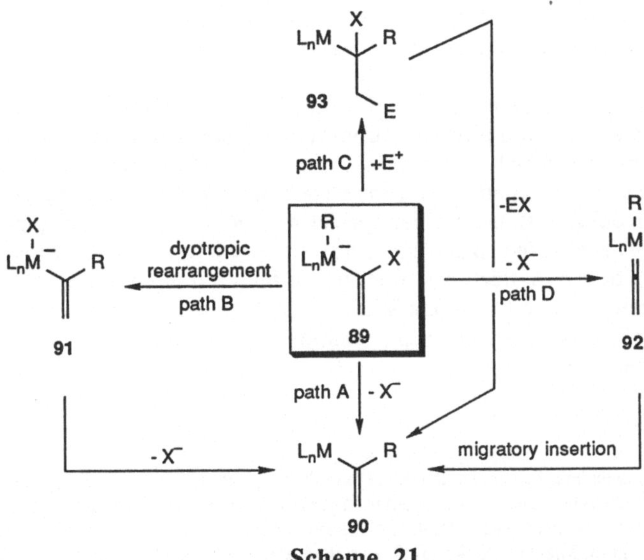

Scheme 21

Path A. Path A provides the simplest and most direct mechanism by which a 1,2-alkyl shift from the metallate onto the α–carbon of the alkene takes place with concomitant displacement of the leaving group and it is the path we invoke in the absence of other information. Since the molecular orbital description of the reaction involves the interaction of a C-M σ-bond (HOMO) with the adjacent C-X σ* (LUMO), the migrating group R must approach the alkenyl centre from the back side with respect to the leaving group X leading to inversion of stereochemistry. We believe that path A explains the TMSCl-assisted rearrangement of α-alkoxyalkenyl borates (section 2), α-(carbamoyloxy)alkenyl borates (section 4) and α-alkoxyalkenyl cuprates (sections 5 and 6).

Path B. Like the mechanism in path A, the dyotropic rearrangement in path B is a concerted process in which the 1,2-alkyl shift from the metallate to the alkenyl carbon is accompanied by displacement of the leaving group X and its simultaneous 1,2-migration to the metal. The stereochemical consequence of a dyotropic rearrangement is inversion of configuration at the α-alkenyl centre. Paths A and B are essentially indistinguishable. Metals which are highly oxyphilic such as B [37] or Zr [38] may provide an additional driving force for rearrangement.

Path C. Path C provides a mechansim which is fundamentally different from paths A and B since it involves electrophilic attack onto the β-carbon of the alkene with concomitant *anti* migration of the alkyl group onto the α-carbon. The initial addition/migration step is likely to be concerted thus giving a *trans* relationship between the migrating alkyl group R and the addended electrophile E. The intermediate **93** then undergoes elimination of the electrophile and the leaving group X in an *anti* fashion leading to net overall inversion of the alkenyl centre. There is a substantial body of evidence (sections 2 and 3) which suggests that the intrinsic nucleophilicity of α-heteroalkenes involving first row heteroatoms is amplified on complexation to metallates of B [4,6,11-13] or Al [10] leading to remarkably easy electrophilic attack with concomitant 1,2-alkyl shifts. Whether this reactivity translates to metallate complexes of Cu or other transition metallates remains to be proven.

Path D. The mechanism depicted in Scheme 21 as path D requires as the first step the loss of the leaving group X assisted by a pair of electrons from the metal resulting in the formation of an alkylidene carbenoid **92** which then undergoes migratory insertion to generate the alkenylmetal **90**. Unless the ligands L_n attached to the metal impose a chiral environment or there is some co-ordination effect to an unspecified species prior to loss of the leaving group X, it is hard to explain why there should be any inherent stereoselectivity in the migration of the alkyl group R. The loss of stereoselectivity observed in the rearrangement of α-(carbamoyloxy)alkenyl cuprates (section 7) along with the formation of alkynes, presumably *via* Fritsch-Buttenberg-Wiechell rearrangement, point to the intermediacy of a metal carbenoid if not a free alkylidene carbene.

Acknowledgements

The author would like to express his sincere thanks to the following research collaborators whose experimental skill, patience and perseverance have contributed to the unravelling of the 1,2-metallate rearrangement story: Christopher Barber, Simon Birkinshaw, Paul Bury, Sharon Casson, Nicholas Dixon, Georges Hareau, Michael O'Shea, Austen Pimm, Michael Stocks and Sjoerd Wadman. We are grateful for financial support from Glaxo Group Research, Pfizer Central Research, Fisons Pharmaceuticals, and the Science and Engineering Research Council. We would also like to thank Prof. Dieter Hoppe for advice.

References

[1] D. S. Matteson, R. W. H. Mah J. Am. Chem. Soc. 85 (1963) 2599; Review: *α-Haloboronic Ester Intermediates for Stereodirected Synthesis*, D. S. Matteson, Chem. Rev. 89 (1989) 1535
[2] G. Zweifel, H. Arzoumanian J. Am. Chem. Soc. 89 (1967) 5086; G. Zweifel, R. P. Fisher, J. T. Snow, C. C. Whitney, *ibid.* 93 (1971) 6209.
[3] P. Binger, Angew. Chem. Int. Ed. Engl. 6 (1967) 84; P. Binger, G. Benedikt, G. W. Rosenmund, R. Köster, Justus Liebigs Ann. Chem. 717 (1968) 21
[4] A. B. Levy, S. J. Schwartz, Tetrahedron Lett. (1976) 2201; A. B. Levy, S. J. Schwartz, N. Wilson, B. Christie, J. Organomet. Chem.156 (1976) 123
[5] J. A. Soderquist, I. Rivera, Tetrahedron Lett. 30 (1989) 3919
[6] S. Birkinshaw, P. Kocieński, Tetrahedron Lett. 32 (1991) 6961
[7] E. Negishi, Comprehensive Organometallic Chemistry, G. Wilkinson, F. G. A. Stone, E. Abel (Eds.), 7 (1982) 303; A. Pelter, K. Smith, H. C. Brown, *Borane Reagents*, Academic Press, London, 1988; D. S. Matteson, Tetrahedron 45 (1989) 1859; A. Suzuki, Accts. Chem. Res. 15 (1982) 178
[8] K. Uchida, K. Utimoto, H. Nozaki, J. Org. Chem. 41 (1976) 2941; E. J. Corey, W. L. Seibel, Tetrahedron Lett. 27 (1986) 905, 909
[9] A. Suzuki, Pure Appl. Chem. 57 (1985) 1749

[11] A. B. Levy, Tetrahedron Lett. (1979) 4021

[12] G. Hareau, Southampton University, 1992, unpublished results

[13] M. Ishikura, M. Terashima, K. Okamura, T. Date, J. Chem. Soc. Chem. Commun. (1991) 1219

[14] For a summary of the stereoselective synthesis and reactions of enol carbamates see: D. Hoppe, Angew. Chem. Int. Ed. Engl. 23 (1984) 932

[15] For the first example of the metallation of enol carbamates see: P. Kocieński, N. J. Dixon, Synlett (1989) 52; for further extensions see S. Sengupta, V. Snieckus, J. Org. Chem. 55 (1990) 5680

[16] M. T. Reetz, Adv. Organomet. Chem. 16 (1977) 33

[17] R Paul, S. Tchelitcheff, Compt. Rend. Acad. Sci. Paris, 235 (1952) 1227

[18] C. M. Hill, G. W. Senter, L. Haynes, M. E. Hill, J. Am. Chem. Soc. 76 (1954) 4538

[19] F. L. M. Pattison, R. E. A. Dear, Can. J. Chem. 41 (1963) 2600. For a recent re-investigation of this work see: E. Negishi, T. Nguyen, Tetrahedron Lett. 32 (1991) 5903; see also ref 36.

[20] M. Stähle, J. Hartmann, M. Schlosser, Helv. Chim. Acta 60 (1977) 1730

[21] T. Fujisawa, Y. Kurita, M. Kawashima, T. Sato, Chem. Lett. (1982) 1641

[22] P. Kocieński, S. N. Wadman, K. Cooper, J. Am. Chem. Soc. 111 (1989) 2363

[23] S. H. Bertz, J. Am. Chem. Soc. 112 (1990) 4031

[24] For proof that higher order cyanocuprates do exist see: B. H. Lipshutz, S. Sharma, E. L. Ellsworth, J. Am. Chem. Soc. 112 (1990) 4032

[25] M. Ochai, K. Oshima, Y. Masaki, J. Am. Chem. Soc. 112 (1990) 7059

[26] W. F. Bailey, E. R. Punzalan, J. Org. Chem. 55 (1990) 5404; E. Negishi, D. R. Swanson, C. J. Rousset, J. Org. Chem. 55 (1990) 5406

[27] R. K. Boeckman, K. J. Bruza, Tetrahedron, 37 (1981) 3997

[28] Reviews: *Applications of Higher Order Mixed Organocuprates to Organic Synthesis*, B. H. Lipshutz, Synthesis (1987) 325; *The Evolution of Higher Order Cyanocuprates*, B. H. Lipshutz, Synlett (1990) 119

[29] S. H. Bertz, J. Am. Chem. Soc. 113 (1991) 5470; S. H. Bertz, G. Dabbagh, Tetrahedron 45 (1990) 425; G. H. Posner, Org. React., 19 (1972) 1

[30] P. Knochel, N. Jeong, M. J. Rozema, M. C. P. Yeh, J. Am. Chem. Soc. 111 (1990) 6474

[31] M. Stocks, P. Kocieński, D. K. Donald, Tetrahedron Lett. 31 (1990) 1637

[32] C. Barber, P. Bury, P. Kocieński, M. O'Shea, J. Chem. Soc. Chem. Commun. (1991) 1595

[33] P. Fritsch, Justus Liebig's Annalen der Chemie 279 (1894) 324; W. P. Buttenberg, *ibid.*, 279 (1894) 337; H. Wiechell, *ibid.*, 279 (1894) 337; G. Köbrich, Angew. Chem. Int. Ed. Engl. 4 (1965) 49

[34] J. R. Behling, K. A. Babiak, J. S. Ng, A. L. Campbell, R. Moretti, M. Koerner, B. H. Lipshutz, J. Am. Chem. Soc. 110 (1988) 2641

[35] M. M. Olmstead, P. P. Power, J. Am. Chem. Soc. 112 (1990) 8008

[36] For a review of the 1,2-metallate rearrangements in saturated as well as unsaturated systems see: P. Kocieński, C. Barber, Pure Appl. Chem. 62 (1990) 1933

[37] A. Suzuki, N. Miyaura, M. Itoh, Tetrahedron 27 (1971) 2775

[38] G. Erker, R. Petrenz J. Chem. Soc. Chem. Commun. (1989) 345S